"A spellbinding serial voyage in which encounters with islands across time are gathered, displayed and reburnished. Memoir becomes morality, as the oldest human myths challenge present neglect and political malfunction."

Iain Sinclair

"Illuminating, incisive and beautifully written."

Kirsty Young

"From ancient Crete to modern Canvey, this is a fascinating voyage around island identity, exploring isolation and imagination through a wealth of stories from around the world."

Martha Kearney

"A timely and original exploration of the liminalities of islands and the waters that envelop them: by turns beguiling, enchanting and ultimately affirming."

Sir Anthony Seldon

"This is a huge theme which Mark Easton pursues with vigorous and beautifully clear prose. His archipelagic fascination is contagious. Read this and the maps in your mind will never be quite the same again."

Peter Hennessy

ISLANDS

Searching for truth on the shoreline

MARK EASTON

Biteback Publishing

First published in Great Britain in 2022 by
Biteback Publishing Ltd, London
Copyright © Mark Easton 2022

Excerpt from 'Sea Fever' by John Masefield reproduced by kind permission of the Society
of Authors as the literary representative of the estate of John Masefield.

Excerpt from 'Nation's Ode to the Coast' by John Cooper Clarke reproduced by kind permission
of Edge Street Live and the National Trust.

Excerpt from 'The Schooner *Flight*' by Derek Walcott reproduced by kind permission
of the Derek Walcott Estate and Faber & Faber.

Excerpt from 'Exodus' by Bob Marley reproduced by kind permission of
Fifty-Six Hope Road Music Limited/Primary Wave/Blue Mountain (Irish Town Songs ASCAP).

Every reasonable effort has been made to trace copyright holders of material
reproduced in this book, but if any have been inadvertently overlooked the
publisher would be glad to hear from them.

ISBN 978-1-78590-776-0

10 9 8 7 6 5 4 3 2 1

A CIP catalogue record for this book is available from the British Library.

Set in Freight Text

Printed and bound in Great Britain by
CPI Group (UK) Ltd, Croydon CR0 4YY

To my family

The Sleeping Lady, a Neolithic statuette found in the Hypogeum,
an underground burial chamber in Malta.

CONTENTS

FORESHORE

Isolation, *n. 'the state of being unhappily alone'.*

The word 'isolation' travels with bags packed full of negative connotations. Prisoners are kept in isolation as a punishment. Isolated people are seen as a cause for concern. 'Isolation is the sum-total of wretchedness to man,' the historian Thomas Carlyle suggested in 1843.

But during the pandemic of 2020–22, social isolation became a goal, a public health target, a national and global objective. Those who defied the rules on isolation faced retribution and condemnation.

Humans (*Homo sapiens*) are social creatures. In moments of joy, we congregate and embrace. When we are fearful, we hold hands or place a comforting arm around a troubled shoulder. We console those dear to us with a hug or a gentle kiss. But Covid-19 corrupted those qualities of communion and connection for its own destructive ends. It crept into our world by exploiting the warmth of humanity itself. The breath in our lungs. The touch of our fingers. The love in our hearts.

Rather than coming together, we were instructed to keep our distance. Instead of touching, we were told to wash away all traces of physical contact with soap and disinfectant. Coronavirus forced us to contradict all that our instinct was screaming for us to do.

The pandemic tested the relationship between the individual and the community, the singular and the plural. We were advised to see our family, friends and neighbours as potential killers, unwitting hosts for an alien predator that was waiting to pounce. When the virus patrols allowed, we were ordered to skirt around those we encountered in the street, ensuring we maintained social distancing at all times. Our personal sovereignty had to be protected from the invasive menace. It felt discourteous and unnatural, but we were required to observe the formal etiquette of the epidemic.

Humanity may be a synonym for compassion and generosity, but it is also part of human nature to build walls around us, to pull up the drawbridge in the face of danger. The virus encouraged us to cocoon ourselves from risk, to dig a moat and cut ourselves off from an external threat. Governments closed borders and grounded flights. Towns and cities set up roadblocks and barricades. Families stocked up on vital supplies and shut the door on the outside world.

In the end, of course, it was humanity that overcame Covid-19, the healing power of kindness and cooperation, of resourcefulness and courage. Distancing reminded us of the importance of togetherness. Physical boundaries, walls and barricades could not prevent new social connections being made or stop love entering our hearts.

Having recently emerged from our island caves, I think this is a good moment to assess the special place where isolation meets connectedness, to go mudlarking upon the shorelines where 'us' meets 'them'.

CHAPTER 1

ISLANDNESS

The Magic Circle

I always have a sense of where the sea is. I can feel its protective embrace around me. I am an islander. Perhaps it was those childhood summer holidays spent on the island of Arran (*Eilean Arainn*) in the Clyde estuary, paddling tentatively in the rock pools, wincing as barnacles and limpets scratched my feet, skidding on the bladder wrack draped across the middle shore. I loved to explore this precarious territory shared by both land and ocean, the intertidal zone where boundaries are fuzzy, neither one thing nor the other. Some days we would head out in a small boat to fish for mackerel around the Pladda Island lighthouse, staring at the mysterious uninhabited Ailsa Craig (*Creag Ealasaid*) on the horizon and glancing back to the beach from where we had departed.

My professional and personal lives have taken me to all kind of islands: from studying the concentrated Englishness of Canvey Island in Essex to the marvellous mind maps of the Mer people in the Torres Straits; from communism in Cuba to capitalism in Singapore; the cruelty of Robben Island's cells and the remarkable happiness of Hebridean villagers; post-colonial tensions in Mauritius and post-war identity politics on the Falklands; religious retreats and tax havens; dumping grounds of the poor and luxurious hideaways of the rich.

More than forty years in journalism has seen me trying to stand further and further back from the events that power daily news, seeking perspective and context. And in doing that, I have headed for the extremities, the outliers, the people and places on the edge of the scatter graph. That is where one finds clarity. Often, they are islands, physically or metaphorically, where a particular behaviour has been contained and intensified.

It is what we saw when the enforced isolation of Covid-19 turned up the contrast on personal traits and national characteristics. The emergency brought out the best and worst in people, exceptional deeds of kindness and appalling acts of cruelty behind the door of domestic confinement. As governments sought to turn their territories into island fortresses, the restrictions magnified cultural differences in responding to moments of crisis, between societies generally motivated by the 'common good' and those with a liberal tradition focusing on the interests of the individual. The statistics reveal a marked contrast in the impact the virus had on populations in the east (where Confucian, Buddhist and Taoist communitarian values are prevalent) and the west (where European liberalism tends to hold sway).

This phenomenon was particularly apparent among island

nations. On 11 March 2022, the second anniversary of the pandemic being declared, islands which had seen the lowest death rates were predominantly in the South Pacific and the Far East, while almost all the twenty island states with the highest death rates were in the Caribbean and Europe. Proportionately, Trinidad and Tobago, the United Kingdom, the British Virgin Islands and Bermuda suffered forty times as many deaths as Taiwan, ten times as many as Japan.

I remember being despatched to Tokyo for the BBC in the early '90s and being quite amazed by what I encountered. As a correspondent who had reported on a decade of Thatcherism and the social strife that went with it, I thought I had a reasonable grasp of how societies function, the motivations that drive people's behaviour. Disembarking in a country which, like Britain, was famous for its 'island mentality', I realised the individualism I had naively assumed was an inevitable driving force in social politics was not inevitable at all.

If you want to get a better understanding of the human condition in all its impossible complexity, islands are good places to go. Through their stories and experiences, they help unravel what is going on and identify the underlying factors that shape the actions that hit the headlines.

Islands have a psychological effect on those who live within the outline of their coasts, a strength and a vulnerability, a sense of exceptionalism but also of disconnection. My country, Britain, is proud of its island status. Tradition and heritage, a way of life framed and protected by white cliffs and rocky shores, are central to the character of the nation. But Britain is also a country that wants to reach beyond its boundaries, to be part of a bigger conversation, open to new ideas and embracing change, taking the risks that keep an island from ossifying and stagnating.

That contradiction in the character of a country, between looking inwards and looking outwards, is the source of the storm whipped up by the recent debates over Brexit in the United Kingdom. It helps explain the furious waves that have crashed, not just on my own island's shore, but on communities and nations around the world. It is about the contested nature of islandness, where nationalism meets globalism, local meets universal, inside meets outside.

Wanting to test that sense of British islandness, I find a proxy for it using Hansard, the parliamentary record of every word uttered in proceedings at the Palace of Westminster. I search for the phrase 'island nation', from the earliest records in 1800 to the present day. It turns out the expression was not uttered by any parliamentarian in the entire nineteenth century, when Britain saw itself as builder and commander of a global empire. The first occasion was in 1904, when an MP referred to Britain as an island nation of island people, 'lying as we do between the old world and the new'. Perhaps its use marked a moment of anxiety, a niggling worry that the tectonic plates of continental power were shifting along the mid-Atlantic ridge.

But the phrase didn't catch on. It cropped up fewer than fifty times in the next seventy years. There was a brief flurry in the early 1980s as MPs and peers discussed the Falklands

War and the decline of the UK's once dominant shipbuilding industry. But if the language of the Houses of Parliament can be used as a measure of Britain's sense of its islandness, then it has recently reached levels never previously seen. The phrase was deployed more times in the three years after the EU referendum than in the first 180 years of Hansard's record.

During my coverage of the Brexit campaign for the BBC, I decided to film at Hever Castle in Kent. It is the fortress of every child's imagining, high castellated walls above a deep moat, the only access by a drawbridge. A metaphor for the country and its relationship with the lands beyond the water, it also symbolises something else in the country's psyche, the idea that an Englishman's home is his castle, territory controlled by the landowner, connected to but distinct from the wider world.

'Should we pull up the drawbridge?' I asked startled visitors. It was a question about islandness, about insularity and isolation, but also about connection. People can feel safer when the drawbridge is raised, protected from external threat and alien influence. But as the occupants of Hever Castle once knew, if the drawbridge is up, it also increases other risks, the danger of being besieged, of being cut off. The castle is an island within an island.

How people define themselves, the exclusive or inclusive nature of their identity, is a facet of islandness. We all draw lines around ourselves, deciding who we are and who we are not, concentric circles of loyalty and belonging, islands within islands.

My ambition in this book is to offer a better understanding of the almost magical qualities of the encircling shoreline, both physically and psychologically.

'No man is an island, entire of itself,' John Donne famously declared. Rather, he explained, we are all a piece of the continent, a part of the main. I hesitate to challenge the great scholar, but this is to misrepresent the nature of islands. They are not closed systems, unaffected by the outside world, entire of themselves. Rather than no man is an island, I would argue we are all islands in that we are individuals, bound by the attributes of our bodies and our minds, but shaped by all that washes up on our shores, the experiences and influences that come from the outside, a product of nature and of nurture.

The tidal zone, territory contested by earth and water, is a rich but hazardous environment. In people, it is the same. The place where personal sovereignty and privacy flirts with alliance and exposure thrills and terrifies in equal measure. Islands have a powerful hold upon us because we see in them the boundary between our own individuality and the wider world. They demand each one of us to answer the question, 'Who am I?'

Every island, like every person, is shaped by its separations and connections. Islands are a product of what happens at the junction between land and sea, upon the beaches and beneath the cliffs, what the tides throw up and what they take away. That interaction is unique to each island and the stories of how that process has played out allows us to examine the human condition, from the most profound personal emotions to the building blocks of society and the natural world. Islands are Petri dishes in which we can study how cultures bloom and colonies grow.

Biologists refer to them as natural laboratories, not because islands are cut off from the outside world, they never are, but because their relationship with the wider world is uniquely distorted by their isolation. 'Island syndrome' is the term used to describe the effect of insular separation on the morphology, ecology, physiology and behaviour of species, describing how isolation magnifies certain elements and shrinks others. In the natural world one might think of odd-sized island dwellers like the giant tortoise and Komodo dragon, or the miniature Sumatran rhinoceros and diminutive Shetland pony. It is the same with human societies – islands allow the development and survival of exaggerated cultural and political systems.

At the individual level, island thinking can amplify the good and bad, the virtues and vices of each one of us. Academics have, of course, come up with a simple acronym for it: ABC – amplification by compression. By distorting the conventional, islands help us understand how the world functions and how human beings behave.

Over the past twenty years or so, there has been an academic movement pledged to changing the way we think about islands. There are complaints of a 'continentalist' view of history in a world that often regards islands as peripheral and inconsequential, side-lining their experiences and stories to the margins, considering them as subordinate and needy.

Instead, it is argued, we should join the dots, to draw a chart revealing the cross-currents and connections, thinking *with* the archipelago as it is described.

It is tempting to support the underdog in this argument, to cheer for the little islands over the big continental bullies. But that is to miss a more important point. If one looks at the world and humankind through an island prism, piecing together the experiences and stories of the shoreline, one may get a little closer to the truth.

That is the challenge I have set myself, to chronicle the journey of physical islands and explore the psychological islands that form the great archipelago of humankind. To that end, there are two books forced together in one, a risky enterprise for which I ask your indulgence. I want us to explore the place where the random spray of subjective personal reflection crashes against the rocks of objective documented historical fact. As I have described, the shoreline is contradictory and dangerous, the place where 'I' meets 'us' and 'us' encounters 'them'. And so we begin, by leaving home and heading out across the ocean, to visit the first island of all.

CHAPTER 2

PANGAEA

Finding Our Island Mother

Alfred Wegener was many things: Arctic explorer, record-breaking balloonist, astronomer, kite flyer and weather forecaster. But he was not a geologist. At least, he did not have any formal geology qualifications. What he did have was a curious and open mind.

When a work colleague showed Alfred the atlas he had received for Christmas in 1910, the young German meteorologist noted how the pregnant bulge of South America would fit rather snugly into the welcoming curves of West Africa. It was hardly an original observation, but Alfred's enquiring personality saw him seek out evidence that the two continents may have once been joined and somehow drifted apart.

He discovered remarkable similarities between plant and animal fossils found in Brazil and Gabon, between marsupials in Australia and in Argentina, between layered rock formations at the edge of one continent and another many thousands of miles apart. Alfred began having ideas. Scientific theorising was a dangerous occupation, his father-in-law warned him, particularly for an outsider and especially someone not granted residency on the exclusive island of qualified geologists. But Alfred was not to be dissuaded.

The symposium of the American Association of Petroleum Geologists in New York in 1926 was described as 'spirited', an adjective only rarely applied to gatherings within the desiccated, dusty world of geology. On the agenda was Alfred's short paper, setting out his theory on the origin of continents and oceans.

Many of the men in the room were advocates of permanentism, a belief in the primordial nature of land and sea. Terra firma was essentially the same now as it always had been, designed, some believed, by God. Each continent, like their thinking, was fixed. Alfred represented an impertinent challenge to their certainties. It wasn't simply that he lacked geological training; he was also a German who had fought with some distinction in the Great War. On the other side.

Alfred's paper suggested that all land on the planet was once joined together in an 'urkontinent', a huge landmass formed some 200 to 250 million years ago. He had given his supercontinent the name Pangaea, a Latinised version of two Greek words: pan (all, entire, whole) and gaia (Mother Earth, land).

One professor stood up to dismiss the hypothesis of

continental drift as footloose, 'less tied down by awkward, ugly facts' than rival theories. 'Facts are facts,' another said, 'and it is from facts that we make our generalisations, from the little to the great, and it is wrong for a stranger to the facts he handles to generalise from them.'

Alfred would not live to experience vindication. On an Arctic expedition in 1930, he died while attempting to deliver food to his camp. His body, at the request of his widow, still lies in northern Greenland, upon territory now regarded as the earth's largest island, but once, of course, an integral part of Pangaea.

The theory of continental drift evolved into the now accepted science of plate tectonics, but Alfred had been a disruptor, breathing new life into fossilised thinking about the nature of the world. He was a reminder of how fresh ideas often came from outsiders and how putting up impervious walls could ossify everything within.

Islands come, and islands go. It is reckoned that, currently, there are several million of them dotted around the planet, from the smallest uninhabited rocky protuberances that barely count, to vast landmasses like Greenland and New Guinea which flirt with being too continental to claim the honour of island status. A physical island is defined by the water that surrounds it, literally and semantically. The quality one might call islandness, therefore, must be a property of that junction: where land meets sea, where detached bumps into connected, where independence encounters alliance.

As with islands, the character of a person is defined by behaviour in the space where the individual meets the wider world. I have an impression of me in my head, a sense of who I am, but such internal notions are almost meaningless unless tested externally. It is at our own personal shoreline that we find out who we are. Likewise, the character of societies is a consequence of how they interact with those beyond their borders, a relationship most obviously displayed among island peoples.

To get a better understanding of islandness and the effect of island syndrome, one must go to islands and meet islanders. Answers will be found at the frontier. I need to walk the coastal path, become a beachcomber seeking clues among tidewrack and flotsam, paddling in the shallows, immersing myself in the surf. But where, among the countless islands, should my journey start? As a small boy, I was taken to see the film *Doctor Dolittle*, and I vividly remember how the title character chose to begin his quest by randomly sticking a pin in a map, ending up on the tropical Sea Star Island. I decide to adopt an only marginally more systematic approach, seeking to identify the median island as my starting point, the island that is right in the middle of all other islands.

I take down my atlas and open it at the page marked the Mediterranean Sea, the middle sea that is home to more than 300 named islands, an alphabet of insularity from Ada Bojana in Montenegro to Zvërnec in Albania. Then I place my ruler carefully across the whole page, with Tangier at one end and Beirut at the other. Halfway along the ruler's edge, bang in the middle of the middle sea with a name that sits precisely and pleasingly in the middle of the alphabet is a small archipelago of eight islands. My journey will begin in Malta. Right in the middle.

The story of all the islands we know could be said to begin with rumbling birth spasms on the supercontinent of Pangaea some 200 million years ago. Tremors and tsunamis, earthquakes and volcanic ruptures provided the violent soundtrack as landmasses tore apart and drifted on their way.

North America, Europe and Africa sought personal space as the Indian subcontinent divorced itself from Antarctica, headed north by north-west and smashed into Asia, throwing up Himalayan peaks. As this slow seismic salsa played out across the expanse of aeons, a vast constellation of islands was formed, some only to be consumed again by continents or drowned by oceans, others to secure permanent isolation.

Fifty million years ago, the islands of Madagascar and Mauritius had already escaped from their continental parent. Colliding plates along the Pacific rim of fire had given birth to the Philippines. Greenland and a vast colony of islands in the Canadian arctic had achieved autonomy. New Zealand had broken free from Australian shackles and headed for a remote corner.

As the great geological clock ticked on, coral atolls and microcontinental archipelagos, skerries and stacks, sandbanks and volcanic domes, islands of all shapes and sizes dotted the globe. Some were fragments of the mainland, ripped from their mother with DNA intact. Most were born barren, isolated and alone.

But nature quickly located her lost children and planted seeds on their vacant coasts, little pods of life that could hitch a ride on the ocean currents and trade winds. Creatures of the sea and the air found their way across the shoreline to multiply upon virgin territory. Giant storms and waves occasionally threw bewildered land animals onto these strange new states, each arrival contributing to an independent habitat with a unique character and personality.

After a dusty bus ride from the airport to the centre of Valletta, I walk to Malta's National Museum of Archaeology, an elaborate baroque building in Republic Street. It tells the stories of the Maltese Islands' various settlers and invaders in a series of coloured rooms: the Bronze Age room full of daggers and ancient fragments is decorated in a rusty brown; the hall dedicated to the seafaring Phoenicians is painted in Tyrian purple, the colour of the dye they first produced from the secretions of predatory sea snails (*Bolinus brandaris*); there is a room dedicated to coins, from early Carthaginian currency to a sparkling set of £5 golden coins of British origin. I wander through the exhibition, piecing together the timeline of the island, until I reach the room dedicated to its first settlers, the walls coloured in the creamy yellow of Malta's limestone. Large lumps of this rock, rescued from Neolithic sites, have been placed on low plinths and it is possible to run your fingers along the remarkable spiral and geometric decorations carved upon them by artists from the Stone Age.

In the last room, however, I encounter an exhibit that stops me short. In a glass case, given pride of place, is a statuette labelled the Sleeping Lady. Crafted out of clay around 5,000 years ago, it is a breathtakingly beautiful depiction of an amply proportioned woman asleep on a low sagging couch. The sensitivity and compassion of the Neolithic artist who sculpted her has spilled down the millennia intact.

I am transfixed by this gift from the Stone Age and feel almost as though I have been deliberately led to this small artefact, an object which might somehow provide inspiration in understanding the nature of our islandness. I buy a replica in the souvenir shop and pop it in my bag, resolving to awaken the sleeping lady. Later, as I place her upon my dressing table, I give her the name Pangaea.

CHAPTER 3

SEEDS

The First Islanders

When the first human species evolved around 5 or 6 million years ago, islands visible from the continent's edge must have seemed tantalising but unattainable. Changing sea levels, seismic events and glacial action tweaked coastlines but also described the limits of exploration and settlement. It had been thought that our ancient ancestors only developed the technology to cross water and occupy islands, simple dug-out canoes and rafts, around 70,000 years ago. But then, in 2014, something extraordinary turned up on the island of Luzon in the northern Philippines to make us rethink the island story.

Laurence Wilson, a mining prospector from Nebraska, had arrived in the Philippines in 1930, where he successfully located a valuable source of chromium. But Laurence was also mesmerised by the islands, their peoples and their stories. He loved to search for clues about the tribes who had originally settled on the Philippines, organising fossil-hunting expeditions and writing several books on the lives and legends of the mountain Ilocano, Igorot and Aeta peoples who have retained their ancient customs and traditions in remote districts of Luzon Island to this day.

After his death in 1961, Laurence's papers and journals ended up in the library of the archaeological department at the University of the Philippines, where they lay pretty much untouched for fifty years until a young archaeologist from France, Thomas Ingicco, began leafing through them.

Thomas and an international team of prehistorians were looking for evidence of early human occupation on the Philippines and hoped that a greater understanding of the animals that once existed on the islands would tell them more about where prehistoric people lived and what they ate for lunch. The word 'rhinoceros' leapt out from one of the pages, along with a hand-drawn map showing where Laurence claimed to have seen teeth and bones of a species that was assumed to have died out long before humans arrived.

Thomas and his team headed for the site close to the town of Rizal in the Cagayan Valley and within a week of digging in the thick mud, sure enough, they unearthed a fossilised rhinoceros tooth. The excited archaeologists excavated further and encountered a leg bone and other fragments. Thomas realised that Laurence had led them to a complete prehistoric rhinoceros skeleton. Carefully removing the soil, they looked at a rib still resting in the clay... and their world stood still. Upon the rib were the unmistakeable marks of butchery.

In that moment, everything we thought we knew about when our ancestors first settled on islands was called into question. The bones were tested using uranium-series dating and electron-spin resonance which concluded the rhinoceros was roaming the island between 631,000 and 777,000 years ago. The cuts and bashes to its rib bone, signs of having been stripped of its meat and its marrow extracted, could only have been done by people. Yet Luzon had been an island long before the first apes began walking on two legs.

The evidence from the muddy pit in the Cagayan Valley was that hominids didn't complete the first perilous journey across the ocean 70,000 years ago but 700,000 years ago.

After publishing his findings in the scientific journal *Nature* in 2017, Thomas became something of an archaeological celebrity. He told *Esquire* magazine, 'You have mysteries, then you have theories. Even if you find the answer, more mysteries will come out. How did they arrive? That's the new mystery.'

How did they arrive, but also why did they leave? The never-ending search for food must have encouraged early humans to seek fresh pasture, but it was surely the human thirst for answers and a natural spirit of adventure that saw them develop the technology to make ocean boundaries permeable and islands accessible. The world beyond the horizon was a mystery that had to be solved, whatever the risk.

In the last great Ice Age, lower sea levels meant that current islands including Borneo, Sumatra and Java were connected to the south-east Asian peninsular, part of continental Sundaland, as it would later be called. Australia bulged out to gobble up Tasmania and Papua New Guinea along with other smaller islands, a landmass named Sahul. Such land-bridges may explain how the rhinoceros arrived in Luzon. They also allowed some ancient peoples simply to walk to newly accessible areas before the oceans swept back, isolating them from their ancestral roots. The rising waters would not have washed away the primeval yearning to reconnect with a homeland still visible, and seafaring must have been the logical solution to unlocking emotional and spiritual reunion with their cultural birthplace.

Island-hopping – that is, sailing from one visible island to the next – led early humans along archipelagos and island chains, fostering a spirit of exploration that developed into the next stage of settlement, defined as 'island-hoping': heading over the horizon, destination unknown. Simple rafts were improved with dug-out logs for buoyancy, evolving over time into canoes with one or two outriggers giving the vessels stability in the ocean swell. The first boats capable of long-distance deep-water travel were fashioned by intrepid and inventive tribespeople in the Pacific, probably around 50,000 years ago.

These courageous seafarers would point their prow towards the skyline and set sail, voyages of discovery inspired by blind optimism. They were on a mission to seek out new life and new civilisations, to boldly go where no man had gone before. Far off islands, to early man, were the equivalent of distant galaxies to the crew of the starship *Enterprise* in the TV show *Star Trek*.

Among the earliest of these Stone Age Captain Kirks were the remarkable Lapita people, who set sail without chart or compass from what is now Taiwan, heading south along the Philippines, then east along the coast of New Guinea, past the Solomon Islands and on out into the deep ocean waters of remote Oceania. They could not have known of the existence of the Santa Cruz archipelago, of the islands of Vanuatu, Fiji, Samoa and Tonga. But eventually they found them, pulled their sailing canoes onto the sand and sowed the seeds of a new culture, some 5,000 years ago.

Despite the vast distances and treacherous seas, the Lapita moved their families and domesticated animals out to the new communities, travelling back and forth many times. They brought plants to cultivate and other supplies in clay pots. It must have been like building a city on the moon. The Lapita, though, were reaching for the stars.

Linguists had long puzzled over the striking similarities between the ancient Paiwan language, still spoken on the southern coasts of Taiwan, and Malagasy, the language of Madagascar, 5,500 miles to the west off the coast of Africa. The word for the number one in Paiwan was '*ita*'. In Malagasy it was '*isa*'. The number nine was '*siva*' in Paiwan and '*sivy*' in Malagasy. Both languages were described as Austronesian, in the same linguistic family as the indigenous island tongues of Hawaii, Fiji, Tonga, the Solomon Islands and Samoa. It seemed the Lapita people headed west. A long way west.

In large outrigger canoes, guided by ocean currents, clouds, sea birds and stars, these extraordinary sailors apparently dipped their testicles into the sea to test the temperature and help them navigate vast distances. They were single-minded in their determination to scatter their seed as widely as possible.

Digging deep into Madagascar's soil and history, archaeologists recently extracted charred microscopic remnants of plants to see what the earliest settlers were growing. The answer was rice and mung beans, seeds and pulses that originate in south-east Asia. In Madagascar, the first settlers were known as *vahoaka ntaolo*, a phrase that would have been recognisable in South Borneo and Polynesia. It meant canoe people from the beginning. To this day, the DNA of some Madagascans contains strands that lead back across the Indian Ocean and the South China Sea to the island of Borneo and on to the island of Taiwan.

I take a closer look at Pangaea, asleep on the table. She is wearing a simple pleated skirt that covers the broadest of hips and thighs and is naked from the waist up. Hair falls to her shoulders from the sides of her head, the top of which appears to have been shaved. Her serene face rests upon a pillow held in place by a plump arm, her eyes gently closed.

To the Neolithic artist who shaped her features, she was a supernatural figure of fertility and fecundity from the dawn of time. To me, Pangaea is the mother of all islands, her offspring strewn over the oceans. What might be in the pillow clutched tightly beneath her head? I picture the sack being sliced open, its contents spilling out and scattering upon the winds and the waters. It contains the seeds of life in all shapes and sizes: from microscopic spores of ferns and mosses blown on the breeze, to coconuts (*Cocos nucifera*) and the giant coco de mer (*Lodoicea maldivica*) bobbing optimistically on the waves; rice canoes embarking from their panicle ports; creamy rafts launching from the fat bellies of pumpkins; seeds of sunflower and sweet pea, cumin and fenugreek incubated in the stomachs of migrating birds; precarious craft each with a unique individual aboard, fate unknown.

Gazing out of the bus window as it lurches its way across the Maltese countryside, I wonder at the motivation of the first settlers on this Mediterranean archipelago, the Stone Age people who determined to sail south towards the sun, away from the familiarity of their Sicilian villages, not knowing what lay beyond the horizon. These islands would have been covered with woodland, dense forests of holm oak (*Quercus ilex*) and Aleppo pine (*Pinus halepensis*), the floor beneath the trees a rich and fertile breeding ground for flowers and insects, birds and mammals. A giant swan (*Cygnus falconeri*) and the Maltese hippopotamus (*Hippopotamus melitensis*) used to call this home. When those Neolithic sailors arrived in Malta, it must have been like the Garden of Eden.

We tend to think of island mentality as introverted, isolationist, insular in its most literal sense. But the character of the original islanders was the opposite. They were explorers, adventurers and trailblazers, determined to plant their swashbuckling seed far and wide. Their search for a better place beyond the horizon reflected the optimistic half of human nature, the belief that our role on the planet was a long and hazardous journey that would eventually lead us to paradise.

In 1902, Malta's capital Valletta was alive with the sounds of banging and hammering. The island was home to the Mediterranean fleet, the largest and most prestigious squadron in the British Navy defending the sea link between the United Kingdom and its empire to the east. The dockyards were not big enough for the ten first-class battleships of the Royal Navy to be safely moored, so extra wharves were being constructed. Hundreds of labourers had been brought to the island to assist with the expansion, requiring new accommodation.

In Paola, a commuter town across the harbour from

Valletta, workers were cutting cisterns for one housing development when a hole opened up beneath them, revealing a cavernous space. The builders did what builders have always done and stood around the hole, looking into it and sucking their teeth. Their first instinct was to pretend they had never found it, to cover it up again and carry on as if nothing had happened, but word got out that something very strange had been discovered beneath Paola's scruffy streets.

The hole was a window into an unimagined ancient world. It had revealed a subterranean cave system older than the Pyramids, a Neolithic burial chamber full of thousands of bodies, remains and artefacts which had lain hidden for millennia. Before professional archaeologists could control the site, many of the bones were removed. Objects placed in the graves disappeared before they could be properly catalogued. Vital clues to the story of the Stone Age society which had constructed this labyrinth out of the solid rock were almost certainly lost in the time it took to put scientific systems in place. But the underground burial chamber, the Hypogeum as it was to be called, could not be silenced, and started to give up its secrets.

News of the incredible find spread around the world and within a few years, while the excavations were still going on, visitors were being led into the caves to marvel at what had been uncovered. Forty feet below contemporary street level, the upper part consisted of many rooms, probably extensions to natural caves fashioned 5,000 or 6,000 years ago with crude tools: flint chisels, stone axes and antlers. Some of the rooms had been used initially as places of worship and then as burial chambers, but when those filled up, the builders burrowed down further, hauling the excavated rocks and boulders to the surface. The middle level was more ambitious, a circular main chamber sculpted from solid limestone from which led a series of rooms. Smooth walls had been richly decorated with geometrical spirals, dots and honeycomb patterns painted in red ochre. Amulets, beads, buttons, pots, little stone sculptures of animals and birds and figurines of people were found scattered here, extraordinary objects that were removed to the National Museum of Archaeology for safe keeping. Among those ancient artefacts was the statuette called the Sleeping Lady.

The entrance to one chamber, now named the Holy of Holies, had been chiselled out to resemble a series of trilithons, two large vertical stones with a third lintel across the top, echoing the entrances to Megalithic temples found above ground in Malta. As with those temples, it seems the entrance had been placed at such an angle that it would have been illuminated at winter solstice if the rays of the setting sun could have penetrated.

The lowest level did not contain any bones or significant artefacts and may never have been completed. No one knows why they stopped.

The caves were an international sensation. The king

and queen of England, princes, Presidents and Prime Ministers were given a tour. The Hypogeum appeared on a set of George VI definitive postage stamps, the Maltese 1½d available in slate black, scarlet, blue and green. Another stamp featuring the Hypogeum, issued by the Maltese Post Office in 1980, commemorated the acceptance of the Hypogeum as a UNESCO World Heritage Site.

But the caves were not able to handle celebrity status. All the tourists increased the humidity and carbon dioxide levels in the cave system, causing the growth of green mould which began to cover the surfaces and destroy the unique artwork on the walls. The caves were being eaten alive. Eventually, the Hypogeum was closed to the general public, only a handful of visitors allowed in at a time, the atmosphere constantly monitored by a micro-climate controller.

The architects of this amazing subterranean construction were the descendants of the first settlers on the islands, Neolithic hunters, and farmers from Sicily. At around the time Lapita sailors were driven by an intoxicating desire to leave their footprints upon the sand of unknown and distant islands, these European Stone Age people boarded rafts to make the relatively short crossing to Malta. Once there, they set about extracting huge limestone blocks to build rock temples, the remains of which still dot the landscape and where, it appeared, they made animal sacrifices to a fertility goddess. They had come to worship the earth mother.

CHAPTER 4

UNDERSTANDING

Seeking Answers at the Ends of the Earth

We cherish our individuality, our unique DNA, the parts of ourselves that tell us we are special. But our identity is also about belonging, the traits we share with others. My search for an understanding of human islandness must be conducted in the place where 'I' becomes 'us', the boundary between singular and plural.

I take a short stroll in the Mediterranean sunshine to another of Malta's institutions that promises to tell the island story. The Postal Museum boasts that every stamp ever issued by Malta Post is on display. I feel compelled to go inside.

My paternal grandfather John was, among other things, a philatelist. He took pleasure in collecting and studying stamps of the British Empire, an interest that led him to write a series of well-received books on postage stamp design. I have inherited his fascination with cataloguing and classification. It feels like a confession, but the complete set of Queen Victoria definitives issued by the Maltese Post Office in 1885, from the halfpenny green to the one-shilling violet, delivers a strange shudder of excitement.

The little pieces of gummed paper, in their illuminated cases, are a graphic history of the islands, a record of practical connection with the outside world but also indicative of how the territory sees itself. Commemorative stamps, issued to mark important moments and events, reflect the priorities and passions of the islanders, their social and cultural disposition across time. The character of the place can be found within the perforated frames.

I get a similar thrill from taxonomy, understanding how every organism fits inside one classification box, within another larger one and another and another until eventually you reach a box the size of a kingdom. *Homo sapiens* falls within the genus *Homo*, part of the *Hominidae* family, of the order *Primates*, of the class *Mammalia* in the phylum *Chordata* and ultimately a resident of the kingdom *Animalia*. There is pleasing logic and order to this identification system that helps me visualise the relationships that define the world.

I remember as a young boy writing my address like this:

Mark

Easton

My Bedroom

My House

My Street

Bearsden

Glasgow

Scotland

United Kingdom

Europe

Planet Earth

Solar System

Milky Way

Universe

It was a way of orienting myself, visualising my identity as concentric circles of territorial belonging, a series of islands within islands, beginning with my individual given name and ultimately floating in a sea of infinity. The further one travels, from the 'I' in the middle to the 'us' on the outside, the less one knows. It is a journey to the edge of our understanding.

The Skull Room was the idea of Sir William Turner, the son of a cabinetmaker who went on to become professor of anatomy at Edinburgh University. Thousands of eyeless heads stare from their display cases, accusatory in their fixed grins. Opened in the 1890s to house the medical school's large collection of donated human remains, the elegant two-storey gallery was to become both a place of scientific study and a macabre tourist attraction.

With its neat rows of skulls, sometimes two-deep behind the glass, extending from floor to ceiling, labelled and numbered, this was a place where the search for knowledge and understanding met the souls of men and women. Each exhibit once contained the memories and experiences of a living, breathing human being, many with a grim backstory.

Edinburgh's Anatomical Museum was the product of Victorian fascination with collecting, cataloguing and classifying. During the nineteenth century, well-educated British men took pleasure from amassing large quantities of human bones, artefacts from cultures and tribes around the world. Skulls were particularly popular.

In 2017, the museum removed ten heads from their cases and lined them up for a photograph. The press was summoned to witness the identity parade. For the first time in many centuries, the skulls had spoken.

Reporters were told how scientists had taken the bones from their dusty cabinet and placed them on the spotless benches of a sterile laboratory. Dressed in protective suits, gloves and masks, the research team had then delicately extracted teeth from the heads. The bones, it was explained, all belonged to Guanches, the mysterious aboriginal people of the Canary Islands, the skulls having been found in caves on Tenerife and Gran Canaria. The root tip of each tooth yielded a tiny sample of ancient DNA and, after washing, spinning, extricating, purifying and sequencing, the calcified extract gave up the secrets of the islanders' identities.

The Guanches' arrival on the Canaries, a group of volcanic islands some sixty miles west of the Moroccan coast, had long puzzled anthropologists. It had been assumed that the first settlers had made the journey from North Africa, perhaps having spotted the smoking peak of Mount Teide, a Tenerife volcano, on the distant horizon. But the people of the mainland were Berbers, mountain farmers and herders with little or no experience of boats or seafaring, and the prospect of launching themselves into the wild Atlantic would have been utterly terrifying. Their teeth, though, confirmed that was exactly what they did. One day, roughly 2,500 years ago, a group of intrepid Berbers boarded crude rafts, pushed off from the coast and headed west across the ocean, towards the spot where the smoke rose and the sun set, towards the edge of the world.

According to the BBC's Editorial Values, I must seek the truth. The corporation encourages its journalists to leave the safe and comforting island harbour of accepted wisdom, to search for answers beyond the horizon. There is jeopardy in this adventure because the truth is often uncomfortable. It may well be preferable to stay at home in ignorance than have one's assumptions and prejudices challenged by realities beyond the garden gate.

I must allow the winds of truth to take me where they will, gathering evidence, verifiable pieces of information that can be assessed and assembled to form coherent narratives, recording and explaining the events of the real world. It is binary. On one side there is fact and rationality. On the other lurks rumour, supposition, misinformation, conspiracy and downright lies. My professional mission is to explain what we know and what we don't know, to expose obfuscation, error and untruths. Mysteries are there to be solved, myths to be exploded.

This approach to understanding requires that every question has an answer and suggests human progress should be measured by the accumulation of knowledge, the number of the questions we are able to answer correctly. The meaning of life is a quiz.

As a boy, I liked to read the weekly magazine *Look and Learn*. Each new edition was added to the stack on my bedroom floor, a mountain of facts and pictures and maps and stories that brought their subjects to life. Every copy included a general knowledge quiz, which I liked to do with my father, a bonding ritual that perhaps explains why I have tortured my own children with quizzes and puzzles at almost every opportunity.

I also had a shelf groaning with ring-binders that came free with the first edition of various monthly publications, magazines that turned into a complete set on a particular topic: volcanoes; freshwater fish; the British Empire; postage stamps; space exploration. I can recall the almost physical pleasure I felt when the last magazine slotted into its folder, the final edition that came with an alphabetical index for the entire set.

With Pangaea asleep on the edge of my hotel dressing table

and the Maltese sunshine streaming in through the window, I open my laptop to examine one of the most influential documents in the history of human civilisation. Claudius Ptolemy's map of the world was part of an atlas that came with his extraordinary work *Geography*, originally produced around AD 150 but, having been lost and found, was still being used, adjusted and updated more than 1,500 years later. *Geography* was made up of three sections containing eight books, a complete set building into an academic record of everything that had been identified about the lands and the seas, providing coordinates for all the places and geographical features of the known world. It was, perhaps, the inspiration for those part-work magazines that had so stimulated me as a boy.

Maps are wonderful. I like to run my finger along coasts and mountain ranges, devise zig-zagging routes along roads and over bridges, across oceans and around islands, take a ruler to calculate distances between mysterious places with exotic names, speculate at the human relationships secreted among the shapes and lines and numbers. Ptolemy's map is totally absorbing.

I can see Malta (*Melita*) among dozens of recognisable islands clearly marked in a familiar-looking Mediterranean Sea. As one moves away from the Med, accuracy diminishes. The British Isles are identifiable, albeit twisted and stretched. To the east, a large island in the Indian Ocean represents an oversized Sri Lanka and continuing eastwards the waters are drawn lapping on the Asian coast as far as the South China Sea.

At the westerly edge of the map is a straight line dividing the white of the earth from the blue of what is labelled Oceanus, the name given to the vast, mythical ocean-river that ancient Greeks believed encircled all the land: Panthalassa surrounding Pangaea. To Homer, it was the great sea god who girdles the world, the father of everything, the Titan who guarded the gateway between the dominions of the living and dead.

Following the land–sea boundary south, I spot six small identical islands in a neat line, like a set of buttons on the hem of the world, nestling at the most westward point from the African coast. The map labels them *Insulae Fortunatae* (Fortunate Isles), later to be renamed the Canary Islands. Ptolemy joined the dots of the six little islands and proclaimed the result the prime meridian, the line of longitude defined as zero.

The honour of marking the division between east and west remained with this remote volcanic archipelago for centuries. Gerardus Mercator's famous globe, produced in 1541, drew the meridian precisely through Fuerteventura; in 1634, Cardinal Richelieu sliced the line through Ferro, the westernmost island of the Canaries; and sailors in the eighteenth century were still using Ptolemy's zero to avoid getting lost.

I place my ruler along the meridian and stand up as if to give myself a better sense of perspective. That simple line on the map, I realise, has a deep and complex symbolism. It is the boundary between east and west, between earth and ocean, between life and death, between fact and myth, between the known and the unknown.

For more than two millennia in Europe, the Canary Islands marked the boundary of understanding. Beyond them to the west, there was nothing. A blank. The further you travelled from the middle of the Mediterranean, the hazier the information on the lands and the people who might live there. When you reached the volcanic outcrops of the Canaries, that was the edge. After that, knowledge just ran out.

Shrouded in cloud and smoke, at the limits of human comprehension, the islands on the Atlantic horizon took on a mystical quality for the ancient Greeks. To Homer, they were the Isles of Bliss, the Elysian Fields where righteous mortals might spend immortality in paradise. To the poet Hesiod, they were the Fortunate Isles, where the hard reality of earth yielded to the dark fluidity of the ocean.

Islandness is about our relationship with the unknown. Are we more content protecting the boundaries of our comprehension or striking out from the safety of a familiar port to question what we think we know? I switch on the lamp.

Studying Ptolemy's map, I can see he used the phrase '*Terra Incognita*' to describe regions whose existence had been postulated but were yet to be mapped or explored. He did not mark such places '*Hic Sunt Leones*' ('Here Are Lions') as would appear on later maps, or '*Hic Sunt Dracones*' ('Here Are Dragons') as found on two early sixteenth-century globes. Ptolemy had no wish to emphasise the dangers of the unknown. Quite the opposite. He devised documents and charts that he hoped might aid explorers of the future, intrepid voyagers who would one day attempt to fill in the blanks.

For some, the unknown was a challenge that required a response – humankind had a duty to conquer ignorance. For others, however, it was the order of things that certain matters remained shrouded in mystery. Their path to understanding was strewn with intuition and instinct, magic and wonder.

The prevailing winds of the Mediterranean blew towards the former view since the days of the ancients. Greek thinkers sought to rationalise and explain, just as the guiding lights of the Enlightenment strove to pull away the curtain of blind faith centuries later. Elsewhere around the globe, the dominant philosophy was and is different, a belief that the limits of human capability make knowledge itself an illusion. Language simply cannot capture the ineffable mysteries of existence; truth is an unreachable land beyond our horizon. The unknown, therefore, must be accepted, respected and feared.

Pliny the Elder, the man who wrote the first encyclopaedia, pulled together all that was known about the Canary Islands just over 2,000 years ago. He was a seeker after truth, determined to draw a clear line between fact and fiction, and the islands on the edge must have fascinated him. He wrote of how the place was home to enormous lizards (fossil records show a giant lizard, *Gallotia goliath*, had been a resident) and of 'the multitude of dogs of huge size' that gave the islands and, subsequently, a native yellow songbird (*Serinus canaria*), their names. It is probable that these were not dogs (*Canis familiaris*) but monk seals (*Monachus monachus*), a large colony of which once sunbathed upon the beaches of Lanzarote and Fuerteventura.

Pandemics permitting, today those beaches are often packed with tourists doing the same, and while playing in the sand in 2012, a group of visitors came across pieces of an old pot and a large quantity of snail shells. It is unlikely the discovery was particularly exciting for the holidaymakers, but after archaeologists had seen the finds and dug around in that corner of the beach, the local paper reported that it had turned the history of the archipelago upside down.

In the sands, a research team from the University of La Laguna had found evidence of a Roman kitchen: pieces of crockery, plates and cups, stoves, pots and jars, metal hooks and nails, the bones of domestic animals. The fragments of ceramic, dated to the first century BC or AD, confirming that Romans had lived on the island at least 2,000 years ago. The archaeologists also found the remains of whelks (*Stramonita haemastoma*). A lot of whelks. In all, they counted more than 70,000 empty shells. These explained why the owners of the kitchen had crossed a continent and risked their lives to settle on this volcanic rock in the middle of the wild ocean. The beach had been the site of a processing plant producing a commodity said to be worth more than its weight in silver, if not gold.

The whelk in question is a carnivorous sea snail commonly known as the red-mouthed rock shell or Florida dog winkle. To the Romans, though, it was a source of Tyrian purple, the prized and highly valuable dye used for the coloured stripe on the ceremonial togas of imperial senators and other dignitaries. Pliny's encyclopaedia *Natural History* contains the recipe, which involved a vast number of whelks, the removal of a vein from each one, a cauldron and a lot of boiling and skimming. It was filthy work. The dyers would reek of rotting fish, a stench so unpleasant that the Talmud, setting out Jewish religious law, specifically granted women the right to divorce a husband who chose the trade.

What the find on the beach shows is that, around 2,000 years ago in Europe, the cloud of myth and mystery that had shrouded the Canaries was beginning to lift. Islands that once represented the edge of understanding were being absorbed into what the ancient Greeks called the ecumene (οἰκουμένη). They were becoming part of the known world.

I look at Pangaea and feel moved by this magical slumbering figure. The fact she was found in an underground chamber along with the bodies of thousands of Stone Age people indicates she was afforded a supernatural importance. Her very survival across time suggests a protective reverence, that she was worshipped and loved.

But something strange happened to the people who sculpted the Sleeping Lady. Malta's timeline, so colourfully illustrated in the national museum, contains a small gap, just a few decades, between the Neolithic civilisation that lived on the island for thousands of years, and the arrival of an entirely new race of people in the Bronze Age.

I take another bus ride, this time crossing the divide between the main island of Malta and its sibling Gozo to the north aboard one of the regular ferries. I watch the exhausted landscape drifting lethargically past the window as the driver heads towards the district of Xaghra and the extraordinary Neolithic temples at Ġgantija, a thousand years more ancient than the Pyramids or Stonehenge, built from huge blocks of local limestone before the invention of the wheel.

We don't know what happened to the islands' original settlers. Having constructed such impressive monuments and dug out the remarkable underground cave system, they appear to have abandoned their homes and possessions, their jewellery, their toys, their designs, their tools, their pots, their sculptures, their icons, a treasure trove of prehistoric experiences, and simply disappeared.

The islands' Neolithic ancestors may have worshipped the earth mother, represented her maternal and loving form in giant statues and miniature figurines, paid homage and made sacrifice to her fertile power, but ultimately perhaps, they forgot what she represented. It may have been war or disease that did it for them. They could not isolate themselves from human frailty. Or it may have been that the food ran out and it was time to leave.

I cup my hand around the sleeping Pangaea in my pocket. Like the islands they choose to inhabit, people may become isolated from the wider world. They look inward, not outward, and that leads to complacency and self-obsession. That is the curse of the island. That is the curse of humanity.

CHAPTER 5

CREATION

The Muddled Shoreline of Myth and Truth

If the setting sun on the Isles of Bliss represented the end point for mortal Greeks in the ancient world, the dawn might be said to have been on the Island of Creation – Crete (*Krḗtē*), a huge aircraft carrier of an island anchored towards the eastern edge of the Mediterranean.

It was here, in 2002, that Gerard Gierliński was strolling along a beautiful beach with his girlfriend, enjoying a romantic break. Gerard was a palaeontologist from the Polish Geological Institute and always carried a hammer, a camera and a GPS location tool, even when on holiday. As the couple's footsteps trailed behind them on the soft sands of Crete's western coast, Gerard spotted the indentation of a fossilised footprint in the rocks. His girlfriend sat on a boulder while he took a series of photographs, marking the spot with the GPS. Looking closer at the slab of rock, he realised there were actually a number of footprints, and they were, to his palaeontological eyes, decidedly odd.

The trip came to an end, but those prints on the beach had stamped themselves onto Gerard's consciousness, and for eight years he kept wondering about them. In 2010, he felt compelled to take another look, bought himself a package holiday back to the island and headed to the quiet sandy shore at Trachilos where he (and his now former girlfriend) had ambled in the sun. There they were! Clearly marked by the lengthening shadows of early evening, the fossilised prints had not been left by any dinosaur. They appeared to be human-like. Gerard could see the toes and the ball of the foot – the prints undoubtedly belonged to a creature that walked on two feet.

Back in Poland, Gerard checked the age of the rocks in that part of Crete and confirmed what he had suspected – they were from the Miocene era, which meant the footprints were 5.6 million years old, 2 million years older than the oldest human-like footprints previously found anywhere in the world. The Cretan beach had yielded a truly startling discovery, a find that potentially transformed the timeline of humanity. The accepted story, based on fossils found in Tanzania and Ethiopia, was that our ancestors evolved in Africa roughly 4 to 6 million years ago, migrated into Asia about 2 million years ago and didn't set foot in Europe until around a million years ago.

Gerard called a couple of mates, researchers from the institute, and the palaeontological trio headed back to the beach in November 2012 to check out the story again. They cleaned up the stone slab and found dozens more small but unquestionably human-like footprints. Whatever was there all those years ago had hung around for a while, mooching about on two feet, leaving their trail in the sand.

As Alfred Wegener had learned to his cost in the 1920s, the world of geology changes its mind at glacial speed. Having recruited a small posse of supportive professors from Poland, Sweden, Greece, Britain and the US, Gerard attempted to get the details of his discovery peer-reviewed and published in a geological journal. His backers described the reaction it received as 'ferociously aggressive' and 'savagely hostile'. The most influential people on geological journals tended to be committed Africanists, scientists who had made their careers explaining how the 'cradle of civilisation' was absolutely definitely somewhere upon the vast continent of Africa. After a great many rejection letters, a small but respected publication, the *Proceedings of the Geologists' Association*, agreed to tell the world about the extraordinary footprints found on an island shore. In 2017, a peer-reviewed article appeared: 'Possible hominin footprints from the late Miocene (c. 5.7 Ma) of Crete?'

The question mark hung inconclusively at the end of the title, but experts were now beginning to accept the possibility that very early hominins walked to Crete at a time when it was connected to the mainland. The island was, perhaps, trying to justify its creation-derived name.

Crete appeared to be underlining the point in 2010, when a joint American and Greek archaeological survey team came across an equally stunning find on the island, an arsenal of Stone Age tools that also challenged conventional scientific thinking about the timeline of human endeavour.

'We were flummoxed,' US academic Dr Curtis Runnels later confessed. 'These things were just not supposed to be there.'

The 'there' was 300 feet above sea level, in hills above the resort of Plakias, marine terraces pushed up by tectonic plates, soils that hadn't been on the surface for at least 130,000 years. The 'things' were hundreds of stone objects, hand axes, cleavers and scrapers, made from quartz and chert that dated from the Palaeolithic era. The experts couldn't be sure, but they may have been made by human ancestors as much as 700,000 years ago.

What really made the survey team's heads spin was that Crete had been an island for around 5 million years. The tools were evidence of a significant community of people who must have crossed the water to get to the island, far earlier than ever previously demonstrated.

Like the island of Luzon in the Philippines with its butchered rhinoceros bones, Crete, with its fossilised footprints and peculiar ancient tools, was challenging

conventional wisdom, pushing the story of islands' place
in human affairs back hundreds of thousands of years,
into the mists of prehistory.

Upon her low couch, the sleeping Pangaea is an island of still-
ness among a sea of my notes and papers. I am intrigued by
the importance of islands in creation stories, and her presence somehow guides me to ancient Mesopotamia, the flood-
plains where civilisation is said to have been born, the island
of fertility moated by the mighty Tigris and Euphrates rivers.

It was there in 1893, on the site of the ancient Sumerian city
of Nippur, that the 'Babylonian Expedition' from the University of Pennsylvania unearthed a fragment of a clay tablet. The
shard was among thousands of similar tablets, inscribed with
cuneiform lettering more than 3,000 years ago. It would be
another twenty years before historians had deciphered what
the wedge-shaped markings on the tablet said. To their surprise and delight, the fragment contained an account of how
the world began.

It explained how Anu ruled the sky, Enlil the earth and Enki
the water and how the lesser gods laboured on the land and
maintained the rivers and canals. However, the junior deities
resented their strenuous chores and one day they went on
strike. Looking for a diplomatic way out of the industrial dis-
pute, Enki proposed that humans be created to do the work
instead. And so it was that the mother goddess, whom they

called Mami, was assigned the task of creating the human race. Mami took the red blood of the god of wisdom and mixed it with clay to shape figurines, the tablet explains.

I glance at Pangaea and notice how the souvenir designer has replicated remnants of the original red ochre markings on her body, when her Stone Age sculptor had somehow breathed life into a lump of clay back in the mists of time.

The tablet went on to tell how the gods spat upon each figure in turn and ten months later humans were born, among them Atrahasis, the wise one. But the unforgiving earth god Enlil was unhappy with this new creation and, in the end, he decided to destroy all humankind with a terrible flood. The kind-hearted Enki warned Atrahasis what was about to happen, advising him to pull down his house and build an ark. Sealing the craft with bitumen as instructed, Atrahasis put his family and animals aboard and, sure enough, the storms broke and the floods covered all the land. After seven days, the floods ended and Atrahasis offered sacrifices to the gods for preserving life on the earth.

It is a tale repeated time and time again. It is the epic of Gilgamesh and of the hero Ziusudra in Babylonia, the story of Noah from the book of Genesis, the legend of Deucalion and Pyrrha from ancient Greece.

I am struck by how these creation myths feature the destruction and then the recreation of the boundary between earth and water. The floods, it is often told, are caused by heavenly vengeance, divine disappointment at the behaviour of people. But from catastrophe, life is renewed and refreshed. The oceans are baptismal waters, washing away sin and announcing a perfect new beginning, people born again. In Judaism and Islam, ritual bathing or washing echoes the idea of divine forces cleansing humanity of its faults. The Japanese Shinto religion practises '*misogi*', ritual purification by washing the entire body. In Chinese mythology, the Gun-Yu or Great Flood creates a better society from disaster, and there are similar folktales in cultures around the world.

They are flood myths but also island tales. The moment of salvation comes with a small piece of land emerging from the ocean, the top of a mountain providing a safe harbour, an island of hope amid the watery chaos.

One February day in 1893, Arthur Evans was to be found in the flea markets of Sicily looking for antiquities. He saw some carved sealstones inscribed with mysterious hieroglyphs said to have originated on the island of Crete. His eyes lit up and he bought them on the spot.

At school, Arthur had been described as a boy of powerful original mind. The son of a celebrated archaeologist and geologist, he almost flunked his modern history degree at Oxford because he had little interest in events later than the twelfth century. In the end, they gave him a first.

Arthur was obsessed by the ancient and classical world, and after university travelled extensively in some of the more dangerous parts of a volatile Europe, hunting for Stone Age artefacts and clues to the people who once lived there. An adventurous couple of decades followed, during which he worked as a foreign correspondent for the *Manchester Guardian*, was locked up in Croatia for being a spy and was appointed keeper of Oxford's Ashmolean Museum.

Arthur had developed a fascination for Cretan sealstones, engraved gems used in ancient times to leave an impression in clay or wax. He had been presented with one in recognition for his work at the Ashmolean. There were similar markings on objects in the museum collection. He had bought others from antiquarian dealers in Athens. The strange symbols upon them were quite distinct from Egyptian and Mesopotamian hieroglyphics and Arthur's insatiable curiosity determined that he should find out who had written them and what they said.

The Ottoman Empire which had controlled much of south-east Europe for centuries was dying. As its influence and control shrank, new power struggles ensued, unleashing war and suffering to millions across the region. Arthur, though, also saw an opportunity. The island of Crete, off-limits under Ottoman control, was about to open up, and with questions requiring answers, he determined to be right there when it did.

A combination of guile and resolve saw Arthur purchase an archaeological site on the north side of the island, a flower-covered hill that had yielded some objects bearing very similar symbols to those he had been trying to decipher. He led the first systematic excavation and within a few years Arthur put his name to an article in an academic journal: 'Knossos. Summary Report of the Excavations in 1900'. The first section was simply but stupendously entitled 'The Palace'.

Arthur had uncovered the remains of a city belonging to a previously undocumented civilisation. It was thought that, at its height in around 1500 BC, as many as 100,000 people lived there. The city centred around a vast palace stretching up five storeys, covering three acres, made up of over a thousand interlocking rooms: a throne room and ritual cult centre on the ground floor, a monumental staircase leading up to a series of state rooms, courtyards, bathrooms, flushing lavatories, storerooms, rooms for wine and olive presses, workrooms for craftsmen and other staff, a sophisticated water and drainage system and, nearby, a 400-seat theatre.

Among the possessions found scattered in the ancient palace were obsidian knives and arrowheads, axe and mace heads crafted from colourful stones, crystal spindle whorls and clay spools for clothmaking. The excavation team also unearthed a collection of quite beautiful figurines, sculptural masterpieces depicting a woman naked from the waist up, some with exaggerated breasts and buttocks, a mother goddess smiling serenely, her arms aloft.

The palace walls had been painted with fabulous scenes, many depicting bulls, a creature that appeared to have had a special place in the lives of the people who had lived there.

To Arthur, the maze-like palace and the bull worship could only mean one thing: this extraordinary construction on a Cretan hill was the source for the Greek myth of the Minotaur, the half-man, half-bull creature which dwelt at the centre of a labyrinth built by Daedalus for King Minos of Crete in his palace at Knossos.

The Minoans, as Arthur christened them, were the first great civilisation in Europe, an island people whose influence and industry had been subsumed in and concealed by the mists of Greek mythology. Thousands of clay tablets marked with the strange hieroglyphs were found at the site and later partially decrypted, revealing a sophisticated palatial administration which kept meticulous records of food production, stock levels, financial transactions and military capability. They were accomplished seafarers and wealthy traders, shipping and selling their oil, wine, wool and pots across the region. The Minoans helped develop commercial links between the islands of the eastern Mediterranean, connecting with communities on Rhodes, Sicily, Samos and the islands of the Cyclades.

More than 2,000 years before Arthur had found his palace under the hill, the Greek historian Herodotus wrote of how the great King Minos had established a thalassocracy (sea empire) and how the Cretan model was one the Greeks should look to emulate. It was a story, the father of history conceded, based upon hearsay (*akoē*), oral tradition passed down over centuries which was impossible to verify. But Herodotus recognised the validity of myths and legends retold across time in getting to the truth of things.

Myth and history were to remain intermingled on Crete, the island of creation, as an academic thirst for truth suggested this was indeed home to Europe's first people, first islanders and first civilisation.

The Maltese sightseeing bus snakes its way to a little car park on the top of a hill above the red sands of Ramla il-Ħamra beach on Gozo, where signs point me towards Calypso's Cave. Here, the tour guide explains, is where Odysseus was imprisoned by the sea nymph Calypso as he tried to make it back to his home island of Ithaca.

Odysseus did not actually exist, I remind myself, as the day trippers on the bus complain at how a partial collapse of the cave means it is no longer available for the requisite selfies. Whatever the souvenirs in the Calypso Boutique might suggest, the cave is just a cave, and a less than thrilling one, if truth be told. The great gift of the Enlightenment was to bring rationality and reason to our understanding of the world, placing Homer's *Odyssey* firmly in the fiction section of the library.

But this hard-edged binary construction is an unsatisfactory way to explain humanity's relationship with itself and the universe beyond. The tourists taking photographs of themselves in a dusty car park on Gozo are trying to connect with a story that seems to reach into the soul. The *Odyssey* is a mesmerising tale of love and yearning and steadfastness and hope. Homer's poetry still touches millions of people after almost 3,000 years. Artists have been inspired to create extraordinary works motivated by the agony of Odysseus's island torment. He has become a tangible symbol of human survival and courage. Aristotle argued that poetry is superior to history because its language, rhythm and melody speak to what ought to be true rather than merely what is true.

Myth and history are not opposites, alternative answers to a true or false quiz. Their relationship is more like a mother and child.

I sense Pangaea asleep in the dark of my pocket and pull her out into the Maltese sunshine. Island life, she appears to be telling me, intensifies the connection between myth and history. They are muddled like the sand and sea on a wild shoreline. Whatever the BBC's Editorial Values might say about weighing facts, I wonder if it is in the altered gravity of that intertidal zone that truth might ultimately be found.

CHAPTER 6

SOVEREIGNTY

The Circles Separating Us from Them

The physical geography of ancient Greece was an explosion of land and sea. It was as though a giant amphora had been smashed upon the earth and broken into thousands of fragments, countless islands scattered on the surface of the waters, each with its own personality and characteristics, qualities and defects. The Greeks made sense of their world through the relationship between the rocks upon which they lived and the sea which lapped upon the shoreline. Plato said the people of the Mediterranean were like frogs about a pond, and the myths and stories that shaped their culture play on the uneasy meeting of solid and liquid, of land and sea, of mortality and immortality.

The most influential tale in this regard is, of course, Homer's *Odyssey*, the story of a man's wandering across treacherous waters involving chance encounters on numerous fantastic islands, a succession of obstacles as he attempts to find his way to the safety and familiarity of his own island home, sea-girt Ithaca.

From the rose-fingered early dawn of Circe's bewitching island of Aeaea, to the temptations of the lotus-eaters on an unknown isle, the sirens attempting to lure him onto the rocks of their island shore, the battle with the fearsome Cyclops in an island cave, the sanctuary of Scheria and Calypso's prison on Ogygia, the Mediterranean's islands were painted as a source of threat and trial. They were alien landscapes, foreign and strange staging posts on the long journey home. The Mediterranean was a unifying element in Greek cultural life and its islands were the stepping stones of their civilisation.

When a volcano erupted on what became the island of Santorini in around 1500 BC, it created one of the largest explosions in recorded history, triggering a series of tsunamis and obliterating many human settlements, events which weakened the Minoan civilisation centred upon Crete. The Greek military forces based at Mycenae on the Peloponnese Peninsular could smell opportunity in the sulphurous clouds that poured from the volcano. It is thought they seized the moment and ousted the Minoans as the dominant force in the Aegean. The key to power was to rule the waves and the tide had turned upon the Cretan shore.

Craft of a rudimentary kind had been operating regular services to islands in the Aegean since the Stone Age, coracles and canoes heading for destinations on the isles of Melos, Antiparos and Gyali to procure rare obsidian, a strong black glass-like mineral used to sharpen tools and cut umbilical cords. But as the Greek and Persian empires jostled for supremacy, the islands of the eastern Mediterranean took on a strategic and logistical role at the heart of the most valuable commercial operations on the planet. Their ports and harbours were key nodes in a web of power that vibrated with ferocious competition for trade, territory and cultural domination.

Imperial ambition smashed like towering waves against the cliffs of island independence, countless maritime battles waged upon the beaches of the Aegean over centuries. The proud islands of the Mediterranean experienced invasion, occupation and domination multiple times. Alliances and treaties came and went as each island government attempted to forge, shape and protect its own precious sovereignty.

Melos, the ancient obsidian source situated equidistant between Crete and Athens in the south-west of the Aegean, was a case in point. First settled by tuna fishermen in the Bronze Age, the island became a central location for Cycladic culture, its main city destroyed and rebuilt three times. Then it was the turn of the Minoans who brought their influence to bear upon the island's character, until the shock of the Santorini volcanic eruption allowed the Greek military forces from Mycenae to take control.

By around 1000 BC, a colony of Dorians had established itself on Melos. An ethnic group from the Greek city-state of Sparta, this breakaway community had come to the island wanting to set up their own independent territory, establishing a distinctive Melian way of life.

The islanders issued their own currency, coins bearing the image of an apple, a pun on the island's name and the ancient Greek word for the fruit (*mêlon*). They developed their own alphabet, a unique variant of Greek and Cretan scripts. They attempted to remain broadly neutral in the great imperial power struggles that raged around them.

Efforts at quiet self-determination proved to be in vain, however, when in 416 BC mighty Athens demanded the Melians finance their side in the war against Sparta, and then, when support wasn't forthcoming, sent a huge invasion force to stress the point. The Melians were obdurate in wanting to retain their independence, but the Athenians were equally immovable in their imperial ambition. The superpower used its military might to besiege the tiny island's city until it inevitably fell, at which point the army executed every adult man and sold all the women and children into slavery.

The fateful showdown, described as the Melian Dialogue when recounted by the Athenian historian Thucydides, became a case study in political realism. 'The strong do what they can, and the weak suffer what they must,' as he put it. But the near annihilation of the people of Melos, replaced by a colony of Athenians, was a particularly acute example of the agony of islands. The moat of waters encircling their shore encouraged a sense of individualism and independence, but the same waters brought existential threats to identity and culture, not once but many times.

In 405 BC, with Sparta victorious over Athens at the key naval Battle of Aegospotami, the Spartan general Lysander rewarded Melos by returning some survivors of the Athenian siege back to the island. The Melians had lost their apple-stamped currency, their alphabet and their independence. Their distinctive way of life would soon be absorbed into Hellenism, the mainstream Greek culture which was being eagerly exported to every island and mainland community of the ancient world. It is hard to imagine how those returnees must have felt as they walked up the beach to their former homes.

If you draw a line around a space, it instantly changes the relationship between those on each side of that line. It might be a hedge around a garden, the moat of a castle or an administrative border inscribed upon a map – once the circle is complete, an island is formed, creating 'us' on one side of the line and 'them' on the other. The psychological force of this simple action is remarkable.

It is not only a matter of territorial control or dominion. The act of closing the circle triggers a response that goes to the heart of who we think we are. It is about identity.

Sociologists have been trying to understand the power of the phenomenon for a little over a century, since the Yale professor William Sumner introduced the concept of ethnocentrism in his 1906 book *Folkways*, the tendency for people to believe that the values and standards of their own culture are superior to those of others. 'Loyalty to the group, sacrifice for it, hatred and contempt for outsiders, brotherhood within, warlikeness without – all grow together, common products of the same situation.'

Two world wars later and social science returned to the idea, anxious to understand why, almost uniquely in the animal kingdom, human beings seem predisposed to kill each other in vast numbers. In 1979, the Polish academic Henri Tajfel came up with his 'social identity theory', suggesting that people naturally divide the world into in-groups and out-groups.

It reminds me of that postal address I wrote for myself as a boy, my concentric circles of territorial belonging, from Mark in the middle to the Universe at the edge. I was not just defining 'me' and 'us', I now realise, I was also marking out the territory of 'us' and 'them'.

Social identity theory suggests there are three stages in the way we define ourselves. First, we categorise the world around us, devising a taxonomy that draws circles around islands of identity. We look for similarities and differences: male, female, black, white, Christian, Muslim, British, French, middle class, working class, Arsenal, Tottenham. We then decide which of these islands we inhabit, membership which generally translates into a sense of our personal good fortune at being a citizen of those groups. Finally, we compare our groups with others, seeking confirmation that ours are superior, a process that bolsters our self-esteem.

Physical islands accentuate the sense of 'us' and 'them'. The encircling sea provides a clear demarcation of the in-group and the out-group, intensifying feelings of separateness and belonging. Identity is not only who we are but who we are not.

I am on the car ferry from Gozo back to the main island of Malta, the lounge crowded with commuters making the morning journey to work. It is a trip of around half an hour, enough time to drink a coffee and take in the view of Comino, the third largest island of the archipelago, named after the cumin which once cast its sweaty scent along the top of the limestone cliffs. Comino's position made it popular with pirates who would hide in its deep caves and hidden coves waiting to pounce on boats making the crossing between Malta and Gozo.

As I sip my espresso, I listen to a heated conversation at a nearby table between two young women and a grey-haired man. They are discussing the proposed thirteen-kilometre road tunnel the Maltese Parliament has agreed should be constructed between the two main islands. The women are Gozitan but explain how they must take the boat each day as they travel to and from Valletta for work, expressing delight that the link, first proposed in the late 1960s, is finally to go ahead. But the elderly gentleman cannot agree. He too is a proud Gozitan, but he fears a physical connection will adulterate the character of his beloved island.

A petition to the Maltese Parliament in 2019 reflected the views of Gozo residents like him, describing how the 'precious and unique charm' of their island would be lost for ever if motorists could simply drive there like any other destination programmed into a satnav. 'The tunnel means the destruction of Gozo as we know it,' the petition warned.

Sociologists would no doubt point out that the new tunnel road will cut across the seashore that has encircled Gozo for millions of years, breaching the integrity of the line that has defined its identity.

It is a belief common among island dwellers the world over, that insularity has allowed the development of a way of life superior to those beyond their shore. Gozo sees itself as a relaxed and laid-back retreat, removed from the bustle of neighbouring Malta and spared the full impact of mass tourism. There is a fear, particularly among older residents, that their quiet sanctuary from globalisation will be destroyed when access is made easier, that a noisy tide of visitors will wash away the special character of the place, turning the island into just another Mediterranean strip dedicated to booze-bucket hedonism. As one former politician put it, 'Gozo should not become a carbon copy of Ibiza.'

It ought to be said that a significant majority of Gozitans welcome the plans for a fixed link to Malta, believing it will bring practical and economic benefits to the island. The two young women on the ferry, dressed identically in blue uniforms with their hair neatly pinned, are happy to exchange a bit of Gozitan independence for a simpler commute and greater job security. Some supporters have gone further, arguing that the road will strengthen the broader sovereignty of Malta's archipelago. 'The unification of the national territory is a National Issue, and it is highly unpatriotic of any Maltese to oppose what should be an eminently non-controversial cause,' is how one pro-tunnel campaigner characterised it.

I recall a similarly passionate debate about proposals for a fixed link between the English mainland and the Isle of Wight, the chalk diamond set just a few kilometres from the Hampshire coast, in the Solent Channel. Far from welcoming investment in a new transport connection, there have been numerous campaigns for the island to break its links, to be granted greater autonomy so it might protect its 'peace and quiet life which is virtually extinct on mainland England'. Some expressed the hope that they might one day have their own Parliament, coins and stamps, and one resident, Lauri Say, even wrote a song called 'UDI for IOW'.

> We'll pick up our pitchforks and give them such a fight,
> and declare independence on the Isle of Wight.

It may be a cliché, but many of the island's residents are comforted by the idea that their home is in a time warp, reflecting the less frenetic lifestyle of a pre-industrial bucolic England, a place of nostalgia for the good old days. The idea of a bridge or tunnel linking the island with the mainland was regarded with horror by supporters of the Vectis National Party, a fringe political movement which contested a few elections on a promise 'to protect the island rural culture and way of life'.

By contrast, the Isle of Wight Party was in favour of a road link to the mainland, arguing that the prosperity of residents was 'intrinsically a part of the island's capacity to successfully entertain visitors'.

The island's council has now supported the idea of a feasibility study into what its backers call the Solent Freedom Tunnel, although councillors made it clear that 'for political reasons' they were expressing no view on whether such a tunnel was a good idea. Sovereignty is a loaded word in the UK post-Brexit, and the relationship of an island with territory across the water goes to the heart of the debate about control over domestic culture and identity. For some on the Isle of Wight, as on Gozo, one of the great advantages of island status is the ability to incubate oneself from the outside world.

The experience of Melos was of being a trophy in a bigger game. After the Greeks were ousted, the island became the property of the Romans, then the Byzantines, Arabs, Slavs, Venetians, Turks and, briefly, the Russians. Strategically placed on the south-western tip of the Cyclades, the little island was taken and retaken numerous times, the desperate population shrinking to just a few hundred by the beginning of the nineteenth century. But then, in 1821, the islanders heard the ghostly whispers of their

Hellenic forebears and agreed to fight alongside those seeking Greek independence from Ottoman control, a move that would inspire a positive change in island fortunes.

Perhaps they were prompted by a remarkable event that had occurred just a few months earlier. A farm worker by the name of Yorgos Kentrotas had been gathering stones from a ruined chapel near the Melos village of Trypiti when he noticed a small cave. He looked inside and immediately stepped back, apparently awestruck. Yorgos had stumbled upon a beautiful statue of a woman, a sculpture in two parts, hidden from view for almost 2,000 years. His find was christened *Venus de Milo* (Venus from Melos), a masterpiece of Hellenistic art depicting the goddess Aphrodite, now elevated to one of the most celebrated and recognisable artworks in the world, a symbol of love and beauty, the Queen of Heaven and a goddess of sovereignty. It is thought this Venus originally had one of her arms outstretched. In her hand, experts believed, she had been holding out an apple, emblematic of treasured Melian independence, but now, like the arm itself, lost for ever.

CHAPTER 7

PROGRESS

How Islands Unlocked the World

Half an hour's bus ride west from Valletta is the city of Mdina, a walled medieval fortress that glows apricot and honey in the evening sunlight. I am in the splendid cathedral to see a painting by the local baroque artist Alessio Erardi entitled *The Victorious Count Roger Granting Freedom to the Christian Slaves*. It is a piece of political propaganda: a handsome bearded nobleman in golden armour upon his horse, a red cape blowing behind him, his page at his side carrying a banner of the Madonna and Child and a shield with an escutcheon of red and white squares. The subject is Roger I of Sicily, who conquered Malta in 1091, ousting the Muslim rulers and liberating imprisoned Christians. Island legend has it that he cut up his chequerboard coat of arms to create the Maltese flag. A special mass is still held annually to celebrate this popular hero, regarded in some quarters as the father of Malta. But I am less interested in this fiction of national conception than in the identity of Roger's great-grandfather. His name was Hiallt de Hautville and he was born in Nord-Trøndelag in Norway, son of Ivan Magnusson. Hiallt was a Viking.

Historians disagree about the motivation for Viking voyaging, but a clue is in the name: the word 'Viking' means someone who hides in a *'vik'* ('bay'), pirates who sought wealth and status with exploits of derring-do and extreme violence. Islands were resources to be exploited in their quest for loot and land, the holy monasteries upon Lindisfarne and Iona, Innisfallen and Skellig, were easy raiding targets. Quiet retreats became bloody battlefields, solitude replaced by servitude. Colonies and trading posts were established on almost all the islands of the North Atlantic: Shetland and Orkney, England, Scotland and Ireland, the Faroes and Iceland, a chain of island settlement stretching west to Greenland, Baffin Island and Newfoundland.

Hiallt was born at a time when brave young Norsemen were expected to sail from the fjords and out into the ocean. He ended up in northern France in the early tenth century, among the ruthless Viking force which seized control of Normandy (Land of the North Men) to be used as a base to control the Channel and, in time, seek new conquests.

A descendant of the fearsome Viking chieftain William Longsword (*Vilhjálmr Langaspjót*), also named William, famously headed north and conquered England in 1066. Hiallt's grandson, Robert Guiscard, headed south and conquered southern Italy and Sicily, Palermo falling to his army in 1072. The isle of Corfu across the Ionian Sea was added to his dominion a decade later and finally the island of Kefalonia, where Robert died of dysentery in 1085. His brother Roger, the man unconvincingly depicted in the cathedral painting, carried on the family tradition of island-bagging by ejecting the Saracens from Malta six years later.

Walking out from the dark of the cathedral into a piazza filled with the golden light of early evening, I am struck by how islands may appear insignificant compared with the great continental landmasses that dominate the maps, swept up like debris in the slipstream of imperial expansion, footnotes in history. But an isolated archipelago can be the key to unlocking strategic ambition, a watchtower or a guard-post, the safe harbour or the military command base that swings the balance. Islands may offer a perspective on affairs that continents cannot deliver. The Vikings were not alone in understanding the importance of islands when stitching together an empire.

I notice Pangaea apparently deep in meditative thought tucked beside my water bottle at the bottom of my bag. It is time to resume my journey, briefly dropping in again on the Canary Islands where the relative peace afforded by Atlantic isolation is about to be shattered.

The Guanches, still leading their primeval lives on the Canaries in the fourteenth century, could not have known how forces beyond the horizon were coalescing to impact on their protected island existence. In 1341, dressed in goatskins, their bodies painted in bright colours and carrying stone hunting axes, the aboriginal Guanches must have looked out to sea and wondered about the three ships that had been cruising around their islands for five months. What they couldn't know was that the crew of each state-of-the-art vessel was sizing up the scene, conducting surveys of the inhabitants of every island and drawing detailed maps of the geography. The fleet had been paid for by King Afonso IV of Portugal, who was in expansive mood, looking to reinforce his royal power in the region. Before the ships sailed back to Lisbon, the sailors seized four Guanche natives, Neolithic curiosities to be paraded in front of Portugal's elite on their return.

The sight of these 'island savages' caused a sensation across Europe. People could smell opportunity. For merchants, it was the prospect of slaves to sell. For rulers, it was the prospect of lands to conquer. For the church, it was the prospect of souls to save.

I have made my way to Comino to snorkel in its Blue Lagoon, the middle of the middle islands in the middle of the middle sea. I want to reflect on what I have learned about our relationship with islands and isolation, and where better than floating on the edge, in the timeless margins where earth and water embrace?

Blades of sunlight cut through the warm water as I study the shoals of fish scurrying about in the shallows. A rainbow wrasse (*Coris julis*) glides through a meadow of Neptune grass where the cliffs meet the sea. A jaguar round crab (*Xantho poressa*) shuffles beneath a rock. I adjust my mask to observe a beautiful pink and orange sea slug (*Flabellina affinis*) as it makes its predatory way along the shoreline. They are scenes that have been played out and repeated countless times over hundreds of thousands of years. There is no discernible plot in the story, no development or denouement, just fish doing what fish have always done. It is simply the circle of life.

From ancient days, it was common to think of human existence in the same way: as recurring rhythms, the rising and the setting of the sun, the pulsing of the seasons and the swaying of the tides, a cyclical journey, always returning to where it began. It was as though humanity was walking around a circular track, a coastal path that marked the limits of our comprehension. There were events along the way – battles and sieges, disasters and diseases, births and deaths, harvests good and bad – but earthly progress was the act of placing one righteous foot in front of another. The destination, to many God-fearing Europeans at least, was to be found in another realm beyond the grave, at the day of judgement. Many cultures share the idea that the beginning is also the end, the past is also the future, the Alpha is also the Omega.

In Europe in the Middle Ages, however, the idea of progress as timeless and trendless, a never-ending walk around an island of constancy, was being challenged. Instead of a circle, the new thinking saw progress as linear, a belief that our tomorrows can be better than our yesterdays, that humankind was on a voyage with a glorious destination.

I swim to the beach and sit on the sand, staring across the waters of the sheltered lagoon. Seven centuries earlier, young men had been doing the same, voices of encouragement inspiring them to strike out from the shore, out into the wild ocean and over the horizon. It was the start of a new European age: the Age of Discovery.

By the end of the fourteenth century, King John I of Portugal and his English wife Philippa, the daughter of John of Gaunt, had three sons: Edward (a philosophical character), Peter (the diplomat of the family) and Henry (who would become known as The Navigator). The boys would dedicate their lives to implementing a masterplan for the future of the world, a vision founded upon a belief in the cultural superiority of Christian European society, a conviction that would have a devastating impact upon the fate of many islands and islanders for centuries to come.

Edward, Portugal's future king, was a devout and melancholy soul, a lover of poetry and literature who, towards the end of his life, wrote a reflective book entitled *The Loyal Counsellor*. A wide-ranging philosophical treatise on the values for 'living well' (*'viver bem'*), it includes a moral justification for pursuing colonisation in general and crusade against the Muslims in particular. If the Moors wilfully deny Christianity, 'we and all Catholic lords should wage war to convert their lands to the submission of the Holy Mother Church', he wrote. Edward was heavily influenced by the great classical works of the ancient world, tales of heroism and imperial conquest, of asserting control over the lands and the seas, of the search for treasure and cultural dominance.

The second brother, Peter, was calculating and shrewd. Tall, handsome, blue-eyed with a reddish beard, he was said to possess the physical qualities of his English ancestors. Peter was also blessed with significant charm and diplomacy, attributes he employed with great effect as he toured Europe, drumming up support and resources for the family plan, later described as the 'Atlantic Strategy'. Its aims were impressive: to neutralise the Islamic threat, to spread Christianity around the world and to take control of the global supply chain in gold and spices.

For ten years, Peter wooed the rich and powerful, arranging the marriage of his sister Isabella to the Duke of Burgundy, schmoozing the Florentine bankers Cosimo de' Medici and Palla Strozzi, buttering up the business magnates of the Hanseatic League, personally participating in crusades against the Ottomans on behalf of the Holy Roman Emperor Sigismund, with whom he stayed for two years, and encouraging Pope Martin V to offer private support for the Atlantic Strategy. He even went to

Venice to meet the feared doge, Francesco Foscari, who was fighting almost constant wars with Portugal's allies and, as trading partners of the Ottomans, might have been regarded as the archenemy. Such were Peter's diplomatic skills that he was treated as an honoured guest, showered with presents including a fabulous jewel, a manuscript of the Venetian merchant Marco Polo's travels and, probably most helpfully, a series of maps of Ottoman and Venetian trading routes, along with one of the latest maps of the world. This was A-grade intelligence, handed over without a fight, to be given to Peter's younger brother Henry who was back in Portugal, plotting and planning.

Henry had a sweeping sense of his own destiny, as an adventurer and hero, a conqueror and crusader. He was determined to stamp his imprint upon the world that was opening up before him. Henry's personal mission was 'to prove devotion to God by making the seas navigable'.

The secret strategy was to sail through the Pillars of Hercules at the mouth of the Mediterranean, out into the open Atlantic, swing south to circumnavigate Africa and then round into the heartland of Muslim power via the Indian Ocean. Once there, Henry would wrestle control of land, trade and religious devotion from the 'infidels', while restoring the ancient kingdom of Prester John, the legendary Christian elder whom some regarded as a true descendant of the three Magi, the oriental kings who attended the birth of Jesus.

It was, to say the least, an ambitious plan. The route was almost entirely uncharted, and a deep ocean journey of the kind envisaged was unthinkable in the square-sailed Portuguese barcas that operated on Mediterranean trading routes. Success would require a technical effort of a kind never before witnessed. Henry ordered a series of expeditions to head out into the Atlantic and throw caution to the wind. His most experienced commanders set off, some south down the African coast, others heading west into the unknown, battling ferocious seas and terrifying storms. Their mission was to map the coast but also to chart the vortex of winds and currents that span clockwise around the ocean, the gyre that had tossed countless seafarers to their deaths. In 1419, attempting to use the winds to take them home, two of Henry's captains found themselves powerless against the might of a tempest and had almost resigned themselves to their fate. But then, by extraordinary providence, they found that the storm had somehow delivered them into a safe harbour on a previously unrecorded and uninhabited island in the middle of the Atlantic. Offering thanks to the heavens, they called it Porto Santo, the holy harbour.

The island had not only provided the sailors with salvation; it had also revealed its location to Henry, who immediately claimed it on behalf of the Portuguese crown and demanded further exploration of the area. An expedition the following year revealed another larger forested island to the west and soon both islands had been settled, establishing valuable 'way-stations' for the prince's fleets as he pursued his oceanic ambitions. Madeira (Island of the Wood), as the archipelago became known, was, for Henry, a key to unlocking the mysteries of the world. Its strategic importance had been magnified by the infuriating fact that the rival Spanish kingdom of Castile had largely seized control of the Canary Islands, so diligently mapped by the Portuguese eighty years earlier.

The Guanches had suffered horribly in the years since the fleet of Afonso IV had recorded their presence on the Isles of Bliss, seized a quartet of unhappy 'specimens' and exhibited them back in Portugal as evidence of the uncivilised world that lay beyond the edges of the map. Rival forces from across Europe had descended on the islands: slave traders, merchants, soldiers and missionaries all determined to exploit the resources of the Canaries for their own ends. Stone Age weaponry, however expertly wielded, only got the Guanches so far. In the end, each of the islands fell to the invaders, the last vestiges of a primitive lifestyle eradicated from its incubator, if not from the imagination of European writers and preachers.

Henry had long been troubled by the experience of the Canaries. Not the plight of the Guanches, of course, but that a key Atlantic staging post for implementing his masterplan had been claimed by a rival kingdom. In trying to circumnavigate Africa, the first obstacle was to get past Cape Bojador, the desert headland the Arabs called 'the father of danger' ('*Abu Khatar*'). Many sailors refused to go beyond this point, fearing the strong southerly trade winds would make it impossible ever to return home. They needed safe harbours to use on the journey and, given the unwelcoming nature of the Saharan shoreline, that meant islands. Finding Madeira, a seemingly insignificant outcrop in the middle of nowhere, was therefore a matter of enormous consequence. And when, a few years later, another of Henry's captains was apparently driven by storms into the embrace of a second North Atlantic archipelago, the Azores, the prince could barely have believed his good fortune.

The word 'apparently' (above) was required because of what has been described as an inkwell accident. Details of the original events in the ocean were limited to one source: text on a map drawn in 1439 by the Catalan cartographer Gabriel de Vallseca. Unfortunately, four centuries later an ink blot obscured the surname of the explorer and part of the year his encounter with the Azores was supposed to have taken place. The incident famously occurred when the map was being shown to the novelist George Sand and her lover Frédéric Chopin at their island hideaway on Majorca. The discovery of the Azores was so important to Portugal's national story, however, that questions over what really lay beneath the smudge were disregarded in 1990, when the Portuguese Post Office issued a 100-escudo stamp in orange, red and black, commemorating a sea captain it decided was named Diogo de Silves.

Gabriel's map was a wonderfully graphic depiction of the strategic importance of islands, the newly identified Azores radiating dozens of lines, a sunburst of compass bearings stretching in every direction, offering potential escape routes for sailors brave enough to take on the dangers of the deep. In the Middle Ages, the sea was a chaotic and terrifying environment with great symbolism. The Christian Church often portrayed itself as a ship carrying the faithful through the evil of the unpredictable ocean to ultimate salvation. Medieval mariners believed their lives were in the hands of the Almighty, offering regular prayers and supplications for protection from the monstrous perils of the sea. Islands were far more than useful harbours; they were places of redemption and deliverance.

Henry's pursuit of continental domination would not only rely upon the control of remote islands but also on new technology. He had managed to acquire specialised Arab navigational instruments, quadrants and specialist astrolabes called *balesilhas*. He had obtained, by fair means as well as foul, priceless maps and astronomical charts. But what was really holding him back was the design of his fleet: single-masted cumbersome trading ships that were unable to sail close to the wind. He assembled the finest design talents he could and set them to work in the shipyard at Lagos on Portugal's southern coast, challenging them to come up with a craft that could operate reliably in the deep ocean. They sought inspiration in ships from all over the world: junks from China and Malaya with their sternpost rudders, agile Islamic coasters (*qaribs*), lateen-sailed vessels from the Persian Gulf and the little Portuguese fishing boats (caravels) that braved the Atlantic swell in search of a good catch. From all of this, the design team emerged with a new kind of vessel that would open the door to global exploration: an ocean-going caravel.

Henry, Peter and Edward would not see the kingdom of Prester John restored; their masterplan for world domination would not be realised. But their pursuit of knowledge, their conviction and determination, their sense of entitlement and destiny, changed everything. Not least, for islands.

CHAPTER 8

THRESHOLD

The Thrill of Forbidden Fruit

Human beings are captivated by the contradictions of the shoreline, the frontier where private meets public, safety encounters threat, intimacy becomes exposure. It is at the threshold where existence develops meaning for us. Who do we allow in? How far do we venture out? What should we reveal? What do we keep to ourselves?

Glancing at Pangaea, my island muse meditating on the table, I think about how islandness – that balance between open and closed – shapes our behaviour and our personality. It creates the architecture of everyday life, symbolically and physically.

The coronavirus emergency in 2020 forced most people to review their islandness settings, a strange reconfiguring of our relationship with the wider world. Isolation became a state of grace while socialisation became a problem. It was a time for loners and recluses, introverts and misanthropes, those who preferred a solitary lifestyle and whose islandness levels were already set to the max. For most, though, the restrictions felt unnatural and unsettling. It is too early to know whether the experience of lockdown has resulted in a permanent change to our default islandness settings, politically or culturally.

Psychologists refer to a privacy gradient or a hierarchy of space, from the most intimate and private places to full exposure in the public realm. Everyone has their own privacy settings, and where those are put can be described as a measure of personal islandness.

I consider my own home in England, the layout of the rooms and my routine that moves from intimacy, through the semi-public realm, to exposure and back again. Upstairs, the bedroom and bathroom are strictly private. Downstairs are what estate agents like to refer to as reception rooms, places to entertain family and friends, entry by invitation only. Hospitality, literally friendliness to guests, is about sharing access to personal property – a pot of tea or the kitchen table. The hallway is transitional space, easing visitors from the semi-public to the private realms. From the front door to the garden gate is also a transitional zone from the semi-public to the private, graduating the journey out to the big wide world and an environment where one must expect to encounter complete strangers.

Having left the sanctuary of home, I may walk quickly to my car, pulling the door shut on the public realm to create my own privacy bubble. I am in the driving seat, inhabiting a personal but mobile space, able to traverse the shared environment while retaining a comforting separateness from it. The curious intimacy offered by cars helps explain the honking aggression and road rage that some drivers exhibit, behaviour very different from their conduct when not behind the wheel. They are challenged by the strange fusion of the intensely private and the fully public, a troubling sense that their islandness settings are under threat.

Privacy gradients are steeper or shallower in different situations and cultures. Big cities have different social behaviours than small villages. People in the north of England regard themselves as less insular or standoffish than those in the south, happy to engage in an impromptu conversation with a stranger at the bus stop. The Germans place great emphasis upon personal privacy, famous for their heavy doors which they like to keep firmly closed. The Dutch are renowned for leaving their curtains open at night, allowing passers-by to see into their homes as a sign that they have nothing to hide. The Chinese are dubious about the very concept of personal privacy. The word they use for it, *yinsi* (隱私), carries the connotation of illicit secrets, conspiracies and shame. The Japanese have no specific word for privacy at all, regarding the idea as an alien western concept.

We use physical props to regulate our islandness – windows and doors. Windows allow a modest and regulated level of contact – a view looking out and, if the curtains are open, a chance for others to look in. Doors mark the place where the threshold can be crossed, where outsiders might or might not be welcomed inside. They may be opened, but they may also be shut, the drawbridge of our castle can be lowered or raised.

Human beings are attracted to things they cannot have, and a closed door can be seen as a provocation. Psychologists call it the Forbidden Fruit Effect. We are hard-wired from infancy to be curious and to explore and so what we cannot have becomes more attractive. Toddlers love to escape the confines of their play pen. I must admit to a strange thrill at the thought of walking on the grass when the sign says I must not. A high wall is a cue to jump up to see what's on the other side. It is a familiar human trait. The Book of Genesis tells us how Adam and Eve, naked in the Garden of Eden, had no understanding of personal privacy until they tasted the forbidden fruit from the tree of knowledge. The 'Fall of Man', according to the creation story, was a mortal failure to resist temptation and cross the threshold of prohibition. We don't always like to do what we are told.

The Banda Arc is an immense fold in the earth's crust, sweeping 600 miles in a 180-degree curve, marked with a dotted line of volcanic islands emerging from an ocean 23,000 feet deep. Geologists have long puzzled over the surprising geometric shape produced at the point where three major tectonic plates converge and collide in eastern Indonesia, on the western edge of the so-called Ring of Fire, the explosive Pacific region responsible for 90 per cent of the world's earthquakes. They have marvelled at the rapid birth of the islands rising up from the depths, and they have attempted to get their tectonic modelling to explain this odd semi-circular feature.

Romantics might suspect the earth itself was trying to tell us something, because framed by the arc's embrace was an apparently bottomless sea, and in the middle of that sea was a small group of islands so isolated and insignificant many maps miss them out but which might legitimately claim to have done as much to shape the course of human history as anywhere else on the planet. Men have devoted their fortunes and given their lives to unlock the mysteries of these elusive sprinklings of land in the vastness of the ocean. Empires have been sustained by their secrets. The division of power around the globe from then until now was a consequence of the battles for control over these isolated dots.

Why? In the forests of the Banda Islands, one may find tall evergreen trees with a smell so enchanting that botanists gave the species a scientific classification meaning 'fragrant balm' (*Myristica fragrans*). The perfumed trees, native exclusively to these few islands in the Banda Sea, bear a bell-shaped yellow fruit which, when ripe and opened, reveals a layer of red lace, a crimson cowl draped around a hard, brown kernel. In the middle of the ocean, in the middle of the forest, in the middle of a fruit was the seed that changed the world, a rare and fabulous commodity believed to possess magical properties. It was the nutmeg.

Nutmeg had long been a coveted and expensive

product in Europe, prized as a flavouring, a medicine, a preservative and a status symbol, but the Arab traders who brought it from the east had always refused to reveal exactly where it came from. The veil of secrecy around its source was a quite brilliant marketing strategy. In the absence of hard facts, fantastical stories were told of how the source of nutmeg was a rich and powerful nation hidden somewhere in the Indies, an island protected by fearsome sea monsters and powerful spirits.

It was a myth that spawned a literary genre. The seven voyages of Sinbad the Sailor, for example, featured shipwrecks on a succession of islands where our fortune-seeking hero encountered supernatural beasts, hungry cannibals and madness-inducing herbs. The islands also possessed vast quantities of gold and ivory, valleys carpeted in diamonds, giant rubies and other precious stones. They were morality tales about courage and reward.

Sinbad's journeys contained clear echoes from the stories and accounts of Homer, of Pliny and of Marco Polo, who chronicled the 'Marvels of the World' encountered on his travels as a merchant along the Silk Road in the thirteenth century. Marco Polo told of an oriental island where markets abounded with cloves, galangal, nutmegs and all the other valuable spices and drugs that yielded considerable profit, of island kingdoms where man-eating anthropophagi devoured their captives and of a small island, possibly in the vicinity of the Banda

Sea, that had woods of odiferous trees where all kinds of spicery were to be found.

The fourteenth-century Arab historian Ibn al-Wardi produced a celebrated map of the world that also told the stories of islands occupied by man-eating barbarians, of sea serpents and of fabulous riches. Throughout Europe and the Middle East, the same folktale was being recounted.

In that enchanting misty realm where knowledge and facts blend with legend and myth, the islands of the east rose up in the imagination as a cultural challenge to 'civilised society', their tribes portrayed as naked savages who should be subjugated, tamed and offered salvation. The secrets of the islands were safeguarded by mountainous seas and marine monsters, behind impregnable walls of jagged rocks and precipitous cliffs. They were natural fortresses filled with treasure. In this context, the ultimate prize for an adventurer was to locate and control the source of that most mystical island commodity concealed somewhere in the darkest Orient – the nutmeg.

Sitting on the beach listening to the Mediterranean waves, I become aware of how my breathing matches the hypnotic rhythm of the sea: in and out, in and out. Islandness is based upon opposition and contradiction, looking out and looking in. The ancient Chinese concept of Yin-Yang imagines the universe governed by two competing but complementing principles: an inward energy (feminine, still, dark) and an outward energy (male, active and bright). The familiar Tai Chi symbol illustrates how the Yin and the Yang are interconnected, the two forces embracing and also containing an element of the other, symbolised by the island dots in the monochrome swirls. Watching the sea wrap itself around the rocks on Malta's coast reminds me of that ambiguous affiliation.

Aphrodite (Venus), that most contradicted of divinities – variously worshipped as goddess of love and goddess of war, hermaphroditic with breasts and beard, symbol of both hedonism and modesty – is said to have been born from the bubbling foam of the seashore, rising up upon her scallop shell in Sandro Botticelli's famous depiction. Hesiod's poem *Theogony*, written in around 700 BC and explaining the genealogy of the Greek gods, is the source of the myth. It seems he constructed the story to fit with the pre-existing name of Aphrodite – '*aphrós*' translates as 'seafoam'.

But there is something apt about this conflicted character being created in the beguiling yet treacherous environment of the shoreline, in the effervescence of a frothing sea crashing onto the rocks. She is simultaneously provocative and prim, that mixture of revealing and concealing that has a special sexual allure, playing at the margins of innocence and experience, the private and the public. That is the quintessence of islandness.

I am reminded of a famous photograph of Jackie Kennedy

at the funeral of her husband in 1963, dressed in a neat black Givenchy suit and a heavy black veil. The magazines described it as funereal chic, a combination of private grief and public spectacle. The veil is both a protection, a shield from unwanted social contact at a time of profound personal anguish, and also a declaration, an outward expression of inner feelings. She is vulnerable and strong, modest and glamorous. She is Aphrodite.

As a garment, the veil has long been seen as possessing special properties, simultaneously disguise and display. For brides and for widows, for cultures through the mists of time, it reveals as it hides and hides as it reveals. That paradox is the power the veil has had on human civilisation throughout history. It is both demure and tantalising.

I am in a little shop on Gozo where an elderly woman sells items made from handmade lace. A sign on the counter explains how the traditional patterns were exhibited at London's Great Exhibition in 1851 and copied by factories across England. The shopkeeper looks up from her work and smiles.

Like the veil, lace is also about the interplay between solid and space, material to be looked at and to be looked through. It hints at forbidden pleasure, stitched to the hems of fancy clothing, or used to trim frilly knickers, blurring the boundary between decency and nakedness, chastity and intimacy, virtue and vice.

It is a similar story with the *mashrabiya* of the Arab world, the so-called harem windows. They are famous on Malta and Gozo, architectural echoes from the time before Roger the

Viking ousted the Muslim rulers. Protruding from a main wall, an oriel is enclosed by carved wooden lattices in intricate geometric patterns, grills which allow the occupant to look out but prevent outsiders from seeing in. The windows developed into a feature of Maltese architecture and dozens of beautiful old examples have been granted protected status. Indeed, the MaltaPost Philatelic Bureau issued a set of five stamps in 2007, from the €0.19 to the €1.07, illustrating the development of the islands' balconies.

I glance at Pangaea as I consider the power of the shoreline, the exhilaration to be found on the threshold of knowing and not knowing.

CHAPTER 9

WALLS

The Challenge of Unwelcome Visitors

We all use real and metaphorical doors and windows to regulate our relationship with the wider world, an invitation to enter our private territory or, perhaps, just offer a teasing glimpse. But there is, of course, a third element in the architecture of islandness – walls. These are our sheer cliffs, our coastal defences, the impermeable shield we build to keep the uninvited out.

Watching the Maltese landscape glide past the dusty windows of the bus, I become conscious that my route appears to have entered a maze of stone walls. Once I start to look for them, I notice they are everywhere. These islands seem obsessed by bricklaying, every building and street demarcated by methodically constructed divides, all built from the standard limestone blocks that are cut in vast numbers from countless hidden quarries dug deep into the treeless hills. The same buttery stone is used to form more informal barriers around the fields (*hitan tas-sejjieh*), walls designed to protect the fragile topsoil from invasive winds.

The very essence of the island, its internal geology, has been scooped out and used as barricades against the world. Malta, I discover, even boasts a tourist attraction dedicated to its 'limestone heritage', a celebration of the Maltese Islands' unique stone resource. Visitors can witness displays of the pickaxes and hatchets of quarrymen through the ages, the tools of the artisans who carved out the building blocks of the island's identity. It is a museum of islandness.

It seems that every brick in every wall is made from a chunk of the same sedimentary stone, shaped from the powdered skeletons of countless marine creatures, corals (*Anthozoa*), armoured amoebae (*Foraminifera*) and seashells (*Mollusca*), each of which in a previous age had attempted to protect itself with its own wall of calcium carbonate.

Centuries of invasion and unrest has fostered a conservatism among the Maltese people, nourished by the Roman Catholic Church. Divorce only became lawful in 2011 and abortion remains illegal in all circumstances. The islanders have built walls to try to stop the forces of change, to protect who they are from what they might become.

Just beneath the brow of the hill in Gozo's capital Victoria, I stand on the steps of the Cathedral of the Assumption in the Citadel and look around. In every direction there are high walls constructed from blocks of the island's creamy coralline limestone. I walk up a narrow alley, tall windowless ramparts on either side, climbing steps that lead past reinforced military quarters and impregnable dungeons, until I emerge into the open again, where the scale of the fortifications becomes clearer. The entire hilltop is draped in curtains of brickwork, barricades and battlements, all built from the same honey-yellow stone quarried on the island over centuries and dragged into place. The red and white flag of Malta hangs limply above a tower. It is like being inside a sandcastle, waiting for the tide to come in.

The intoxicating smell of nutmeg somehow equipped this strange island fruit with the power to transform the course of human history. Europe's obsession with control of the spice would prompt a devastating change in the fortunes of many islands and islanders, including isolated communities who knew nothing of nutmeg or the Banda Islands. Our story resumes more than 10,000 miles away in the Caribbean.

In June 1864, records suggest a man with a name that looks suspiciously like a misprint, Mr A. R. Birns, was to be found exploring in a cave on the island of San Salvador in the Bahamas when he came across a wooden stool. Mr Birns packed it up and took it back to his home in Oldham in northern England, wondering who might have secreted such an item. It was clearly an old and treasured object, with four short legs and a polished seat. From between the two front feet protruded a grotesque head, with eyes carved as concentric circles, a small nose with flared nostrils, large ears and a gaping mouth. The following year, Mr Birns decided to donate the stool to the Royal Museum in nearby Salford, which registered it as 'an ancient wooden head-rest carved by the aboriginal Indians of the Bahama Islands about 350 years ago'.

Unfortunately, by the time the stool was subsequently moved to Manchester Museum in 1948, its details had somehow been changed to 'wooden Mexican pillow from St Salvador', an attribution that saw this remarkable item disappear into the caverns of institutional storage. Only in the late 1990s did an astute ethnologist and an eager PhD student realise what had happened, belatedly restoring the '*duho*' (for that was what it was) to its rightful place in the collection.

This tale of European appropriation and carelessness felt apt for an object that provided a rare physical link to an island civilisation that had existed in the Caribbean Sea for up to a thousand years but which had been relegated to little more than a footnote in western history books.

The Taíno people were thought to have arrived in the Caribbean in large ocean-going canoes around 500 BC to 200 BC. They had moved up from the mouth of the Orinoco River in Venezuela, bringing with them a sophisticated culture that grew rapidly on the islands of the Great Antilles. Some estimated that by the late fifteenth century there were 3 million Taíno people just on the island they called *Ayiti* (Land of High Mountains), the place that later became known as Haiti, and there were thriving smaller communities on many of the islands. Each tribe had its own identity defined by the shoreline of its dominion, but within the island, life was communal. Their round thatched houses, clean and well swept, would have been home to twenty people or more, extended families sharing spaces that sat around an open village square, along a river or the coast. The Taíno people regarded land, and the crops and creatures sustained by it, as gifts from the gods, all worthy of proper respect. They did not put walls around property.

The Taínos cultivated a wide range of crops including cassava (*Manihot esculenta*), sweet potatoes (*Ipomoea batatas*) and yams (*Dioscorea*), developed an impressive pharmacopeia of medicines and remedies, produced exquisite pottery, wove intricate belts, fashioned fine jewellery and played a game called *batey* on a rectangular court with a rubber ball. Although they had no written language, the Taínos constructed their way of life upon a developed system of hierarchical government with a profound spiritual underpinning. Taíno mythology held that the sacred mountain contained two caves, one from which their ancestors emerged (*Cacibajagua*), and a second, 'the cave of no importance' ('*Amayau´na*'), from which non-Taínos originated. They worshipped a divine creator called Yaya and twelve other gods venerated in mysterious figurines known as *zemís*. Among these was

Itiba Cahubaba (Great Bleeding Mother), depicted with emaciated arms folded over a bulging belly, representing the suffering of her creation, the living earth.

The bean-shaped island they called Guanahaní was home to an offshoot of the Taíno people, the peaceful Lucayan tribe, their name literally meaning islanders. That was where, probably in the last years of the fifteenth century, a Lucayan shaman reverently placed his ceremonial wooden *duho* into a corner of a cave, unaware that coming over the horizon was a force that would soon obliterate his world.

Christopher Columbus was more adventurer than intellectual, but before he set out on his famous voyage in 1492 to find the spice islands of the Indies, he thought it diligent to read as much as he could of what he might encounter. He made hundreds of annotations in the margins of Pliny's *Natural History* and *The Travels of Marco Polo*, noting how the islands of the Far East harboured wealth beyond imagination, lands rich with gold and exotic spices but also belligerent tribespeople who might fry you for their breakfast.

Christopher was excited, though, as he prepared to leave the heavily fortified port of San Sebastián on the small island of La Gomera in the Canaries that bright September morning. The town had been a stronghold for the Spanish during an uprising by the Guanches a few years earlier, and the massive red-brick tower that dominated the harbour gave a sense of fortitude and resilience that echoed Christopher's mood. In his pocket was a commission from the king of Spain empowering him to 'discover and acquire certain islands and mainland' and install himself as admiral and viceroy and governor therein.

Christopher ordered his ocean-going caravels, *Niña* and *Pinta*, to follow his flagship, a large traditional carrack named *Santa María*, in heading due west from the harbour. He was, of course, convinced that this route would bring him around the globe to the riches of the Orient, a journey of about five weeks. Sure enough, a little over a month later, in the early morning of 12 October, a sailor aboard *Pinta* spotted the white cliffs of an island glowing in the moonlight, precisely as Christopher's calculations suggested they might.

The shoreline of Guanahaní that morning, with warm waters lapping onto the soft sandy beaches of the island's welcoming bays, saw the confused meeting of two quite distinct world views. For the islanders, the peculiar craft contained visitors from the divine *Amayau´na* cave, strangers who should be welcomed and offered respect. For Christopher and his crew, the semi-naked savages looked troublingly similar to the Guanches of the Canary Islands and should be approached with extreme caution.

As it turned out, the remarkable hospitality of the Lucayan people washed away the suspicion of the Europeans. 'They took great delight in pleasing us,' Christopher wrote in his report for the king of Spain. 'They are very gentle and without knowledge of what is evil; nor do they

murder or steal. They love their neighbours as themselves, and they have the sweetest talk in the world, and
are gentle and always laughing. Your Highness may believe that in all the world there can be no better people.'

Christopher's diary reflects bewilderment at encountering a people who had a value system entirely at odds
with everything he had experienced previously. 'They will
give all that they do possess, with good will,' he wrote.
'They do not carry arms or know them. They traded with
us and gave us everything they had, exchanging things
even for bits of broken crockery,' before adding, omi-
nously: 'They should be good servants.'

Within fifty years, the Taíno civilisation had all but
been eradicated. Warfare, enslavement, torture, starva-
tion and mass suicide killed many, but the greatest danger
came from European diseases like smallpox and measles
which wiped out 90 per cent of the natives in one year.
The island of Guanahaní was renamed San Salvador, a
Christian cross and the promise of eternal life planted in
the sand. But across the fragile coral margin that signified
the islanders' territory, death had come in many forms.

Meanwhile, the European intruders searched in vain
for the treasure trove and fabulous spicery they had
convinced themselves were waiting to be claimed in the
name of God and the Crown.

You can almost feel the fear of the people who created this place. I look around at the Citadel's high walls. The power struggles of Europe saw the population of Gozo live in almost constant trepidation for hundreds of years, often besieged for months on end as the cannon balls smashed into their defences. This was the last refuge, their hopeful sanctuary from war, enslavement and premature death. But on many occasions, it proved not to be enough. The walls represented vulnerability, not strength. The islanders adopted a siege mentality, fearful of all outsiders, frightened of the wider world, as they cowered in their bunker.

Today, the Citadel is a proud part of Malta's heritage trail, a tourist attraction trying to tell a totally different story, a tale of island resilience. There is an echo of this version of history in the exhibition at the stamp museum in Valletta. In 1938, Malta's General Post Office issued a 2d definitive, bearing the head of King George VI beside an etching of Gozo's fortifications in a cartouche. The design was used throughout the Second World War, changing colour as postage rates changed, and coming to symbolise the islands' resistance against invasion. There are versions in scarlet, slate-grey and a post-war issue in the colour of yellow limestone over-stamped with the legend 'Self-Government 1947', a sign of a country rediscovering its confidence.

A capacity to endure hardship has become a recognised feature of the nation's character. The George Cross awarded to the people of Malta by George VI, in recognition of their heroism and devotion in holding firm against besieging German forces, is stitched into their flag. The Siege Bell rings gravely every day at noon, over the waters of the Grand Harbour in Valletta, echoing off the towering city walls, commemorating those who died defending the island.

The fortifications around the country's capital were regarded in the Renaissance as reason to designate Valletta an 'ideal city'. The military architecture was hailed as the most majestic anywhere on earth. Major General Whitworth Porter, an officer in the Royal Engineers who supervised the construction of the new dockyard in the 1870s, wrote an homage to the walls entitled *A History of the Fortress of Malta*, describing the island as 'the most powerful artificial fortress in the world'. I wonder if the walls themselves, their obduracy and defensiveness, have somehow entered the soul of the islanders.

The mysterious spice islands of the East Indies not only presented the chance of glory but also promised fortunes and power beyond the dreams of men. The strange trees and plants, found nowhere but in their isolated forests, bore fruit so valuable they invested extraordinary authority and influence on all who touched them on their journey around the world. But Europe found itself at the end of a supply chain that yanked at the continent's sense of its own God-given exceptionalism.

For hundreds of years, the people who held the keys to the Banda Sea were island kingdoms which had battled to win authority over the maritime trade routes. From the seventh century, the Srivijayan Empire was in charge, a Buddhist thalassocracy based on the island of Sumatra that had successfully become the market leader in luxury goods, including exclusive rights to the nutmeg of the Banda Islands. By the thirteenth century, the business had been taken over by the Majapahit Empire, a powerful Hindu kingdom from the island of Java. Once sourced and packaged, the products were moved to ports in India and Sri Lanka where Arab merchants controlled the logistics, shipping the cargo into the Persian Gulf and the Red Sea, or moving it by land along the Silk Road and through the continental gateway of Constantinople. Once in the Mediterranean, the Italian city states of Venice and Genoa organised distribution across Europe, offering a menu of rare spices, herbs, remedies and mind-altering drugs, including opium (*Lachryma papaveris*).

The arrangement was suffocatingly frustrating to the Christian kings of the west. Vast profits from the trade were being syphoned off before they reached Europe, wealth that was funding an Islamic Golden Age and the expansion of the Ottoman Empire. The closely guarded secrets of uncivilised islanders upon unknown islands in an uncharted sea were thwarting the righteous aspirations of a continent.

Picking up the baton of Henry the Navigator's mission 'to prove devotion to God by making the seas navigable', a succession of sincere and fearless European men boarded boats in a quest to find the forests of the nutmeg. As we have seen, Christopher Columbus, with Spanish patronage, had mistakenly headed west in 1492. Six years later, and flying the flag of Portugal, Vasco da Gama sailed south and around the Cape of Good Hope to open up a route to the Indian Ocean. Following Vasco's trail, in 1511 Portugal's foremost military strategist General Afonso de Albuquerque sailed with instructions from King Manuel I to head for the Malacca Strait, through which much of the nutmeg and other spices had to pass. The Europeans were closing in.

'It was to do Our Lord's service that we were brought here,' Afonso wrote upon arrival in the strategic seaway that runs between what became Malaysia and the island of Sumatra. 'By taking Malacca, we would close the Straits so that never again would the Muslims be able to bring their spices by this route,' adding that he was certain 'if this Malacca trade is taken out of their hands, Cairo and Mecca will be completely lost'.

Malacca was a bustling multicultural commercial centre, the gateway for exotic spices, silks and porcelain making their way west. The city fell to Afonso's cannons that summer, giving the Portuguese significant control of the trade route through the narrow strait. The general also hoped it might give him access to knowledge, vital intelligence from sailors who could be persuaded to reveal something about the source of the mysterious nutmeg crop that so obsessed the super-rich of Europe.

His luck was in. A large chart was recovered from a Javanese vessel that Afonso described in a letter to the king as 'the best thing I have ever seen'. Upon it was 'the course your ships must take to the Clove Islands, and where the gold mines lie, and the islands of Java and Banda'. The secrets of the spice trade were being revealed, and by November Afonso had gleaned enough to send three ships off in search of the Banda Islands. Early the following year they found them.

On coming ashore, Afonso's crew realised immediately they were not dealing with man-eating savages protected by sea monsters but shrewd merchants who were part of a sophisticated trading network that extended across the many islands of what would later become Indonesia. The crew negotiated a price before filling their holds with sacks of nutmeg, mace and cloves, the aroma intoxicating as they tried to make sense of what they had achieved.

The Portuguese seemed curiously indifferent to the idea of directly controlling the source of the nutmeg, barely visiting the islands for a dozen years and then deciding to run their nutmeg and mace operations through merchants in Malacca. Afonso was, perhaps, more interested in building an empire than running a farm. It may also have been, though, that having revealed the secrets of the islands, the venture itself had lost its spice. It had been the pursuit of mystery, the lifting of the veil, that drove their ambition. Once the facts had been ascertained, they appeared content to let the peaceful Banda islanders continue with their lives. Others, though, would not be so reticent.

The British began sniffing around. In 1579, Francis Drake sailed to the edge of the Banda Sea aboard the *Golden Hind* and landed on the Maluku Islands, where he was lavishly entertained by the local ruler, Sultan Baabullah, after agreeing to purchase six tons of cloves (*Syzygium aromaticum*). The spice was worth its weight in gold back in Europe, but a good quantity of it had to be thrown overboard when Francis's ship ran aground on a coral reef as he attempted to sail home. The distressing incident emphasised the fabulous profits to be made from the spice trade and further frustrated English merchants who remained at a loss as to how to break the Portuguese monopoly in the islands of the East Indies. And then, in the summer of 1592, they had their answer. It came, not from the warm waters gently lapping on white sands on the other side of the world, but from the treacherous seas of the North Atlantic, where waves smashed onto the basalt cliffs of the Azores.

A squadron of English ships, with a commission from Queen Elizabeth I and under the ultimate command of Sir Walter Raleigh, was hiding in the coves of the archipelago, lying in wait for Spanish or Portuguese galleons heading home, their bellies hopefully laden with gold and other booty from the New World. On 3 August, lookouts spotted a ship coming their way, and as it got closer, excitement rose. The vessel was enormous, three times larger than any in the entire English Navy. But, before the sun rose the following morning, the splintered decks of the Portuguese carrack *Madre de Deus* were awash with blood and strewn with bodies. The English had managed to take the ship with her cargo still intact. They entered the hold and jaws dropped. The vast cavern beneath the waterline was filled floor to ceiling with treasure: chests crammed with jewels, pearls, silver and gold; bales of precious silks and calicos; Chinese porcelain and exotic carpets. Most of the space, though, was given over to huge quantities of spices. An inventory counted 425 tons of black pepper, forty-five tons of cloves, thirty-five tons of cinnamon and three tons of the rarest nutmeg. The total value of the consignment, in today's money, would have been around a quarter of a billion dollars.

And yet, the most significant item aboard the *Madre de Deus* was not one of obvious monetary value. According to the contemporary reports of English geographer Richard Hakluyt, it was found 'enclosed in a case of sweet cedar-wood, and lapped up in almost an hundred fold in fine Calicut-cloth, as though it had been some incomparable jewel'. Inside the box was a document its Portuguese owners had prayed would never get into the wrong hands, top-secret papers that would shift the balance of global power for ever.

Richard Hakluyt recorded that God Himself had bestowed upon England 'those secret trades & Indian riches, which hitherto lay strangely hidden, and cunningly concealed from us'. The mysteries of the spice market, the locations of the islands and charts of the seas, had been 'brought into the broad light of full and perfect knowledge'.

Crossing the boundary from not knowing to knowing, encountering the new and the strange, is thrilling but can also be troubling. That is the contradiction, the Yin-Yang, at the heart of islandness. Our natural curiosity demands we leave the safety of our familiar harbour and explore, but the further from home we travel, the greater the challenge to our certainties. We are, in social identity terms, entering the territory of the out-group, people who are not like us.

European explorers searching the globe for treasure islands, just as sociology theorists would have predicted, travelled with a colossal trunk of self-belief, an immutable conviction in their own superiority. Their audacity meant doubts were left on the quayside. The channel of human progress, as they saw it, led in a straight line towards a large sign with the word 'civilisation' etched upon it, and as representatives of the blessed group at the pinnacle of the social hierarchy, it was their historical destiny to tame the savage and claim their reward.

Echoing the adventure stories of Odysseus and Sinbad, reports of terrifying island people returned with the sailors. Western consciousness bolstered its sense of righteousness with these tales of the fearful 'other', wild beasts that looked human but were devoid of morals or learning, prone to cannibalism and human sacrifice, primitive and inferior. Like the unfortunate Guanches before them, islanders from the Orient were seized as exotic specimens and brought back to Europe to be exhibited like animals at royal courts and public fairs. The conclusion appeared self-evident: these creatures lacked the prerequisites of civilised society, and it was therefore the Christian duty of heroic European adventurers to bring religious, political and economic order to such backward souls.

We prefer the island boundaries we draw around ourselves to be unambiguous, white cliffs plunging into a black sea. Strong identities are based upon a clear separation between us and them, between the home and away teams. To increase the contrast, we emphasise the perceived positives of our own in-group and the negatives we identify in the alien out-group. Stories and sermons, tall tales and fake news – all are used to reinforce the sense of our own superiority and the inferiority of those who do not belong to our group. It makes us feel good about ourselves, but without challenge, keen rivalry drifts into hostility. Sociologists describe this process as 'othering', a phenomenon amplified when there is also 'spatial marginality'. Think ghettos and ganglands. Spatial marginality is about islands of territory, and the island dwellers encountered by the European explorers on their divine mission to civilise the Orient had a line drawn around them, like the edges of a target.

Regulars in the Nag's Head Inn in London's Bishopsgate Street could talk of little other than how the confidential files, now in English custody, might change the fortunes of the nation. The pub's Great Room was the meeting place for ambitious merchants who had watched in alarm as Dutch rivals had formed themselves into a company and were already challenging the Portuguese for control of trade with the Far East. The foam of good English ale on their lips, the assembled clientele agreed they should follow on the Dutchmen's tail, calling themselves the Adventurers and deciding to petition the queen to give them monopoly trading rights with all 'islands and countries' in the Orient.

On New Year's Eve 1600, the Adventurers were given their Royal Commission, rebranded themselves as the East India Company and sent a fleet to the Indies under the command of one of their founder members, James Lancaster. Using the intelligence discovered on the *Madre de Deus*, James navigated to the strategic Nicobar Islands, headed south-west to the islands of Sumatra and Java, trying to tie up business deals with the local sultans as he went. In February 1603, his ships now loaded with a fortune in peppercorns, James set course for home, but not before ordering a small scouting boat to head for the clove-producing islands of Ternate and Tidore, part of the Maluku archipelago, then being contested by Portuguese and Dutch business interests.

It is a reminder of how serendipity can have far-reaching consequences that, in adverse winds, the little pinnace ended up failing to reach its destination but succeeding in founding an empire.

On 3 March, its crew found refuge from the storms on two tiny, isolated coral-ringed atolls, the most inaccessible outcrops of the Banda Islands. The smell of nutmeg hung heavy on the breeze as they planted English flags in the sand of Run Island and nearby Ai Island, claiming them as colonies of the Crown. The locals must have been slightly baffled. Their extreme isolation meant they had never owed allegiance to anyone. They didn't have many visitors and their affairs were run, as they always had been, by an elected village council led by a headman (*orang kaya*). It was a way of life built on consensus and respect. The bedraggled sailors seemed unthreatening, and with time, a good business relationship developed into a firm friendship. After a few more years, the headman agreed to accept King James I of England as his sovereign in return for British protection from the feared Dutch, whose vessels menaced their coasts. A ceremony was arranged at which an oath of allegiance was sworn, and the English were presented with a nutmeg seedling, a hugely symbolic gesture. Nutmeg specimens escaping the Banda Sea threatened the monopoly the islanders had on the spice trade. The gift represented the most closely guarded secret of all. King James, meanwhile, was delighted to describe himself as 'King of England, Scotland, Ireland, France, Puloway [Ai Island] and Puloroon [Run Island]'. The British Empire had officially begun.

I am at Malta's International Airport, preparing to head back to my home in London. At the coffee shop, I pull Pangaea from my bag and place her on the table. I detect a note of sadness in her otherwise impassive expression. As her story has unfolded across history, I realise how the strengths of Pangaea's island-children are also their vulnerabilities. Isolation offers strategic advantage that avaricious powers will battle to control. Insularity nurtures rare commodities that outsiders might covet. Coastal boundaries provide protection but no route for escape. Once an island is inked onto the navigational charts and becomes part of the known world, it must learn how to defend itself.

For centuries, the amiable peoples of the ten volcanic Banda Islands had tended their nutmeg orchards. In this isolated island galaxy, each village had developed its own independent character, creating a series of tiny autonomous states that complicated matters for those looking to profit from a secure and coherent trading network. Business in this distant corner of the planet had required effort and empathy, nurturing personal relationships with the headman and council of each island. It took time for trust and respect to build up, but for those who took the trouble, the rewards could be extraordinary.

Jan Pieterszoon Coen came from the polders of north Holland, where his forebears had claimed land from the sea to bolster continental might. A zealous Calvinist, Jan believed he had divine blessing in all he did, and he was a man in a hurry. As Supreme Commander of the Dutch East India Company (*Vereenigde Oostindische Compagnie* – VOC), his ambition was to build a vast maritime empire that would stretch from India to Japan. The volcanic islands of the Indies would provide the capital and strategic base for global domination, the fire and brimstone of a wrathful God to rain down on those who sought to defy him in his righteous objective.

Jan's approach was backed up by written instructions from the board in Amsterdam: 'We ask your attention for the islands where clove and nutmeg grow and we order you to conquer these islands for the VOC, either through negotiation or violence.'

On 11 March 1621, Jan sent a huge Dutch force to descend on Lontor, the largest of the Banda Islands. Within a day, all the key strongholds and coastal promontories were under his control, the islanders having fled to the hills. Facing an army of 2,000 well-armed troops, the headman feared his people would lose the most precious gift they possessed, the identity that had been nurtured over centuries upon the soil of their beloved island. In return for allowing his people to retain some element of religious and cultural freedom, he agreed to surrender their weapons, destroy their fortifications, hand over a proportion of their nutmeg harvest and sell the remainder only to the Dutch. It was a desperate day for the Bandanese, but matters were about to get a great deal worse.

Hiding in the forests, some of the terrified islanders were involved in skirmishes with the invaders, defiance that Jan used as a pretext for his punishment to begin. Forty-four island leaders were rounded up, tortured and executed, their heads impaled on bamboo stakes and put on public display. Tribal homes and villages were burned to the ground. Within a few months, almost all the 15,000 natives were dead, enslaved or banished from their island. Many starved themselves to death or threw themselves off the cliffs rather than yield to the brutal Dutch forces. The wholesale slaughter of the population drew a reprimand from Jan's company directors in Amsterdam, although he remained in post and continued with his ruthless approach to business, a policy that would help turn the VOC into a capitalist superpower, the first true multinational corporation.

The Banda massacre came to be regarded as a profoundly shameful episode in the history of the Netherlands and an indelible bloodstain on the story of global capitalism. The billions that flowed from the exploitation of the spice islands of the east fuelled the development and expansion of the west. But it was a process founded on genocide.

There was a footnote to the terrible story of the Banda Islands. On 31 July 1667, after a series of expensive and bloody wars, the Dutch and the English signed the Treaty of Breda. Part of the agreement was designed to resolve the situation in the Banda Sea, where the English still claimed sovereignty over tiny Run Island, the last of the nutmeg plantations not in Dutch control. England's negotiators agreed to relinquish their territorial claim in return for another island. Run was swapped for Manhattan.

CHAPTER 10

DESTINY

A Matter of Life and Death

At Heathrow Airport, I am greeted at the UK border by Union Jack framed posters with the legend 'This is GREAT Britain'. The word GREAT is written in large white letters set against red in a blue sky. I remember David Cameron launching the 'Britain Is Great' campaign in New York, promoting UK business and tourism ahead of the 2012 London Olympics. 'There are so many great things about Britain,' the Prime Minister told his audience. 'We want to send out the message loud and proud.' The government notification offered a selection of GREAT attributes. 'There are over 6,000 islands in the British Isles and no place in England is more than 120 kilometres from the coast,' it boasted. Islandness was part of the big sell.

National identities are built upon a sense of specialness, a belief that something exceptional exists within described borders, the 'ethnocentrism' that William Sumner identified at the beginning of the twentieth century. If people share a pride in their country's unique qualities and heritage, that helps to create a valuable sense of shared endeavour. Flags and anthems are designed to bolster citizens' confidence that they are members of a select group, to confirm the belief that their island is better than the other islands.

I catch myself humming a tune as I join the long queue at passport control.

> The English, the English, the English are best
> I wouldn't give tuppence for all of the rest!

It is something my parents introduced to me as a young boy, Michael Flanders and Donald Swann's 'A Song of Patriotic Prejudice', parodying the conceit of nationalism.

> The English are moral, the English are good
> And clever and modest and misunderstood!

Nationalism is social identity theory at work, exaggerating a sense of the in-group's distinct qualities while often casting the out-group in a poor light.

> It's not that they're wicked or naturally bad
> It's knowing they're foreign that makes them so mad!

I look up as I finally reach immigration control to see a huge sign confirming that I am about to cross the UK BORDER. I was among the journalists invited to join the Home Secretary

John Reid as he unveiled the new signage in 2006, a rebranding exercise designed to allay fears that our island's frontiers were not properly protected. Territorial boundaries define a nation geographically, legally and symbolically. The large illuminated letters above my head are telling me that I am entering a separate jurisdiction, a political entity with a degree of self-determination and autonomy, able to set its own rules and decide its own destiny.

Less than two centuries after Christopher Columbus had guided his Spanish paymasters to the gold on the beautiful Caribbean island of Hispaniola, the mines had been exhausted and the native Taíno people had been all but obliterated. Regarded as the Pearl of the Antilles, the island had been shucked by greed, emptied and repurposed as an engine of the expanding Atlantic slave trade. Vast profits were being harvested from the sugar cane and coffee beans that thrived in the fertile soils, but it was wealth that relied upon forced labour shipped from Africa to work the plantations. It meant dehumanising the people employed, seeing slaves as a supply of fresh muscle to replenish a stock constantly denuded by disease and maltreatment. The business model placed violence and cruelty at its heart; the crack of the whip was the sound of money being made.

The French flag now flew in the western part of

Hispaniola island, territory named L'Isle De Saint-Domingue, honouring the humble and ascetic founder of the Dominican order. But the slave masters were an ocean away from both state authority and the attributes of their patron saint. In 1685, evidence reached the royal court in Versailles that workers in the colony were being systematically tortured and mutilated. Louis XIV felt the need to assert his authority and ordered the implementation of *Code Noir*, regulations seeking to control the worst excesses of the plantation system. The legislation banning *'traitements barbares et inhumains'* ('barbaric and inhuman treatment') was issued with the full authority of France's absolute monarch, but the slave masters blithely ignored its protections. The law of the land meant nothing in the land of the lawless.

By the 1780s, Saint-Domingue was the richest island in the Caribbean, producing a sizeable proportion of the world's coffee and sugar. The planters had constructed a society designed solely to maximise their personal fortunes. With around half a million black slaves on the island, a workforce ten times the size of the white colonial community, a policy of divide and rule was introduced to keep control.

It was a rigid caste system. At the top were the landowners known as *'grands blancs'* ('big whites'), often French aristocrats who spent as little time as they could on their mosquito-infested properties. Below them were *'petits blancs'* ('little whites'), Europeans who operated as artisans, shopkeepers and the feared overseers, responsible for keeping discipline on the plantations. Next were *'gens de couleur libres'* ('free people of colour'), largely those of mixed race known as *affranchis* or mulattoes, terms now recognised as offensive. Freed from slavery, these islanders generally had had access to education, with responsibilities at all levels of the business. Some even owned and operated their own plantations. However, *grands blancs* insisted *affranchis* wore only simple clothing without adornment or jewellery, a policy designed to emphasise the plainness and subordinate nature of their status and origins. At the base of the hierarchy were *'les esclaves'* ('slaves'), subdivided further into 'creoles' who had been born in the colony and, at the very bottom of the social pyramid, those who had been captured in Africa and managed to survive the perilous voyage to Saint-Domingue.

The classification system applied to the island did its job. Little whites usually begrudged the privilege of big whites. Free people of colour tended to resent all whites. Slaves, generally reviled, were tormented by bitterness and subjugated by despair. In a climate of mutual loathing, suspicion and duplicity infected the population like the yellow fever. It was just as the masters had intended: virulent distrust coupled with brutal retribution kept a lid on organised rebellion.

Islands, however, are never perfectly insulated, and in the summer of 1789, word reached Saint-Domingue of tumultuous events 5,000 miles away in Paris: centuries of French feudal order had been torn down by a new revolutionary government which asserted that all men were free and were equal. It was unclear to the islanders whether the Declaration of the Rights of Man was intended to apply to slaves or free people of colour in the colonies, but the ripples from social and political upheaval in continental Europe had broken upon the beaches of Hispaniola.

Talk of equality and freedom made plantation owners decidedly uneasy. Some began to wonder if they should seize the moment to break free from state regulation and take total control of island affairs. For both free people of colour and slaves, the radical rhetoric of the republicans stoked the smouldering fires of rebellion. It felt as though the wind had changed.

As a cub reporter, I was taught that sunlight is the best disinfectant. Perhaps it is a conceit, but I do believe that unless journalism shines its beam into authority's dark corners, the spores of corruption are more likely to form and spread. Without scrutiny and challenge, those with power will be tempted to abuse it. History constantly reminds us of the wrongs that can be perpetrated when people are prevented from asking difficult questions or offering an alternative view. The narcissism of dictators flourishes behind high walls.

Island syndrome implies the capacity to 'amplify by compression', the attributes of an island's inhabitants magnified by limiting outside influence. One metaphor is the island as a laboratory Petri dish, a sterile environment where particular bacteria propagate and grow. The dish may be incubating a deadly plague, but, then again, it might turn out to be life-saving vaccine.

It is the experiment that William Golding explores in his 1954 novel *Lord of the Flies*. Will the best or worst of human nature prevail among the boys trying to govern themselves on their deserted island? His gloomy answer, of course, is that the savage wins, a conclusion coloured by the twentieth-century horrors of holocaust and world war.

The question was asked again during the pandemic of 2020, when the lockdown saw a sharp spike in the number of people desperately calling domestic violence helplines. However, isolation also prompted extraordinary acts of compassion and generosity. The enforced retreat created the space for quiet reflection and reassessment. One side effect of the virus proved to be kindness.

While isolated in my home, a tale from the mid-1960s pops up in my news feed, publicity for a new book *Humankind: A Hopeful History*, by the Dutch journalist and historian Rutger Bregman. The papers all choose a variant of the same headline: 'The Real Lord of the Flies'. Rutger has pieced together the true story of a group of six schoolboys who were marooned on the uninhabited Tongan island of 'Ata (Pylstaart Island) in June 1965 after they took a fishing boat and got caught in a storm. The teenagers were castaways for over a year and had been given up for dead when their encampment was finally spotted by a passing mariner. What part of human nature had islandness amplified by compression? It turned out to be community spirit. They each looked out for the others and, unlike the boys in William Golding's book, their teamwork kept the fire going for fifteen long months.

Islands do not make people good or bad. They take what is already there and intensify it. The walls of a cloistered monastery or of a high-security prison may affect the population in very different ways. A Trappist monk from the Welsh holy island of Caldey was revealed as a predatory child abuser. An inmate from the notorious Robben Island jail in South Africa won the Nobel Peace Prize.

Other people are a moderating influence, encouraging us to comply with conventional behaviour and thinking. For good or ill, their presence means we are more likely to go with the flow, to conform to the norms and values of the society around us, the orthodoxy of the group. Isolation is a challenge, requiring us to look inside ourselves for the values and principles that underpin our individuality. Our islandness reveals who we really are.

In the middle of August 1791, former slave and voodoo high priest (*vodou houngan*) Dutty Boukman presided over a secret ceremony in Haiti's mountainous interior.

With priestess (*mambo*) Cécile Fatiman at his side, Dutty made an animal sacrifice, and those attending drank its blood and pledged loyalty to their cause. 'The Good Lord sees all that the white people do,' Dutty told the small congregation. 'This god who is so good demands vengeance! He will direct our hands; he will aid us.'

Pompée Valentin Vastey, a ten-year-old boy of mixed race living in Saint-Domingue at the time, later wrote of the blithe brutality that had spawned the revolution. 'Have they not hung up men with heads downward, drowned them in sacks, crucified them on planks, buried them alive, crushed them in mortars? Have they not forced them to consume excrement?' he wrote. 'Have they not thrown them into boiling cauldrons of cane syrup? Have they not put men and women inside barrels studded with spikes and rolled them down mountainsides into the abyss?'

In the Alligator Woods (*Bois Caïman*) of northern Saint-Domingue, a signal had been sent which would set in motion events to transform the future of Haiti and the future of many island communities across the world.

A week later, after sunset, the priest was back in the forest hideout. A storm rumbled on the horizon, flashes of lightning illuminating the eager faces of thousands of slave representatives who had risked everything to gather among the trees. Dutty and Cécile offered prayers and bade the crowd disperse. The moment of vengeance had arrived.

The shackles of servitude were literally being removed, releasing pent-up fury and hatred at atrocities and inhumanities perpetrated over centuries. Masters and mistresses were dragged from their beds, mutilated and executed. The severed heads of children were placed on pikes and paraded in front of the slave army as it sought to destroy everything in its path. Hundreds of plantations were torched, flames spitting sparks into the night sky as the uprising spread, tens of thousands of slaves and free people of colour sparing nothing and no one in their pursuit of what they imagined would be freedom.

The white planters had long feared such an insurrection and had plans in place. Hastily assembled militias struck back with equal ferocity. Thousands of lives were lost on both sides, but six months after the uprising began, slaves controlled around a third of the island territory. The pressure cooker of Saint-Domingue had finally exploded, and its tremors were felt around the world, nowhere more strongly, of course, than on the other island colonies of the Caribbean.

Islands were a key component of the slave trade. Their insularity provided natural confinement, their harbours allowed strategic access to supply routes, their soils and climate offered fertile growing conditions and their isolation prevented unwanted scrutiny and questions.

In the Indian Ocean, the islands of Zanzibar, Mauritius, Réunion and Seychelles were discreet staging posts for slaves being shipped from East Africa to Saint-Domingue and other parts of the West Indies. Most of the labour, however, was transported directly from West Africa, the slave traders dragging their human cargo across the Atlantic to be unloaded somewhere along the chain of islands draped across the Caribbean Sea: from Trinidad and Tobago in the south, heading north along the Lesser Antilles to Grenada, Barbados, St Lucia, Martinique, Dominica, Guadeloupe and Montserrat; turning north-west at Antigua to reach St Christopher's Island (St Kitts) and its neighbour Nevis, then to Anguilla, the Virgin Islands and the Danish West Indian Islands (US Virgin Islands); continuing on along the Lucayan archipelago to the Bahamas, or taking a more westerly course following the Greater Antilles to tie up on one of the larger islands, at San Juan Bautista (Puerto Rico), Hispaniola, Jamaica or Cuba.

Every dot on the chart, every black volcanic beach and pink coral reef described a natural resource that had been seized and devoured by the slave masters. The names on the maps and the flags on the quaysides reflected the ebb and flow of imperial and commercial authority in the region, as ships laden with sugar and coffee, tobacco and indigo, cotton and cocoa and rum headed back across the Atlantic to cities growing tall and fat on their contents. Most people had little or no idea of the grim details of how it worked, but as the eighteenth century drew to a close, the continent of Europe was being powered by distant islands and islanders.

News of the revolution on the isle of Saint-Domingue

set alarm bells ringing in many capitals. In Paris, Madrid and London, urgent meetings were held, envoys briefed, military commanders given their orders and gunboats despatched. A report from the island advised the French government that the only way to retain what it saw as its colonial jewel was to end the slave trade. In 1794, the country's National Convention took the extraordinary step of announcing that 'slavery of the blacks is abolished in all the colonies'. Across the channel, the British were aghast. The Prime Minister, William Pitt the Younger, demanded the largest fleet his navy had ever assembled – more than 30,000 men and 200 ships – set sail to conquer Saint-Domingue and restore order. Half the British Army was deployed to the West Indies.

It is conceivable, one could suppose, that the Prime Minister had been spared some of the gory details of life on the slave islands of the Caribbean. He may have been unaware how yellow fever killed half of new arrivals within a year. He might not have known how treachery and duplicity were embedded in the culture. Perhaps his advisers failed to mention that these were social cauldrons where promises and principles were worth little, where honour and decency evaporated like dew in the tropical heat.

Whether he was properly briefed or not, William's demand for a 'Great Push' was the order of a man who had not taken the trouble to comprehend the vile realities of the trade that powered his country's economy. He would learn his lesson too late.

Allegiances and alliances came and went like the tides on the shore. It was hard to keep up with who was your friend and your enemy, while the greatest threat of all came from the swarms of mosquitoes that sucked the lifeblood out of even the fittest armies. The five-year campaign was a disaster. It had cost the British the equivalent of half a billion pounds, with 100,000 soldiers dead or permanently disabled, when an agreement was finally signed with former slave and commander of the rebel forces General Toussaint Louverture. The deal was simple: the British would leave the island and Toussaint would promise not to support slave rebellions in Jamaica.

With the British out of the way, in 1801 Napoléon Bonaparte sent his brother-in-law Charles Leclerc to Saint-Domingue, secretly planning to restore slavery and national self-respect. The French forces were also ravaged by yellow fever and startled by the ferocity of the slave armies, notably the columns headed by Jean-Jacques Dessalines, which massacred every white person they encountered and piled up the corpses to rot in the tropical sun. Jean-Jacques was a product of the island's regime, born on a sugar-cane plantation but soured by decades of unimaginable cruelty. Toussaint had advised his lieutenant to 'burn and annihilate everything in order that those who have come to reduce us to slavery may have before their eyes the image of the hell which they deserve'. But in May 1802, after an orgy of violence on both sides, Toussaint wearily agreed to a truce: his army would lay down its weapons in return for a French promise that slavery would not be reintroducéd and that he would be free quietly to run his own plantation.

Perhaps it was the stench of duplicity that had long pervaded island affairs, but Napoléon reneged on the deal before the ink was barely dry. Toussaint, the great black general, was seized, packed off to France in shackles and imprisoned in the mountains of Jura, where he contracted pneumonia and died. 'Toussaint, the most unhappy of men!' the poet William Wordsworth lamented on hearing of the soldier's death.

> There's not a breathing of the common wind
> That will forget thee; thou hast great allies;
> Thy friends are exultations, agonies,
> And love, and man's unconquerable mind.

With the armistice broken and their leader gone, the slaves regrouped and fought a guerrilla campaign against the French. Napoléon's commander on the island, Donatien-Marie-Joseph de Vimeur, responded with genocidal vigour. He ordered mass executions of black islanders: thousands were shot, hanged or drowned in bags. For months afterwards, locals refused to consume fish from the bay fearing they might have eaten the bodies of their loved ones. Donatien-Marie-Joseph imported packs of attack dogs from Jamaica, trained to savage only non-whites. The viscount even invented his own method of mass killing, describing it as 'fumigational-sulphurous baths'. Slaves were loaded into the holds of ships and gassed to death using burning sulphur.

If Donatien-Marie-Joseph had assumed he would break the spirit of the island's black population, he was mistaken. When the French executed 500 black people, Jean-Jacques Dessalines executed 500 whites and put their heads on spikes. Previously divided, slaves and free people of colour joined forces in response to the atrocities, tilting military advantage to the islanders. On 18 November 1803, at the Battle of Vertières (*Batay Vètyè*), with a violent tropical storm raging overhead, Donatien-Marie-Joseph was forced to surrender. The French were expelled from Saint-Domingue within ten days. Untrained and largely uneducated islanders had finally defeated the military might of a continent but at huge cost. It was reckoned that the uprising cost the lives of 200,000 black islanders and 100,000 Europeans, one of the most fiercely fought conflicts in history with levels of savagery rarely witnessed.

On New Year's Day 1804, Saint-Domingue declared itself independent, with a new Arawak-derived name – Haiti. The nation's existence presented a series of challenges to western thinking: to the ethics of the business that underpinned its economy, to the belief in white superiority that justified colonialism and to the notion that faraway islands were of little consequence.

CHAPTER 11

PERSPECTIVE

A Different Way of Seeing

I unroll a large poster that once graced the wall of one of my
children's bedrooms but has been housed in a cardboard
tube for many years. It is a struggle to keep it flat, so I weigh
down the edges with a pot of pens on one and the sleeping
Pangaea on another.

It is that most familiar of maps, the world as originally
interpreted by the Flemish cartographer Gerardus Mercator,
with the Americas on the left, the Far East on the right and
Europe, emphasised and enlarged, bang in the middle. This
representation of the layout of land and sea chimed with the
worldview of the rich and powerful of the European continent
in the sixteenth century, the men of influence who had paid
handsomely to own a copy of Gerardus's clever map.

In order to turn a three-dimensional globe into a two-
dimensional chart, Gerardus imagined the planet not as a
sphere but a cylinder, a projection that could be unrolled into
a flat map. It was misleading but practical. Using a grid of

vertical and horizontal lines, navigators could draw a straight route to their destination and set a constant compass bearing. The effect, however, is to shrink land masses closer to the equator and enlarge those nearer the poles.

The methodology persisted and the modern map upon my desk still bears the consequences of what has been criticised as a Eurocentric, even racist, perspective. The island of Madagascar is drawn about the same size as Great Britain when, in reality, it is twice as big. Ellesmere Island (*Umingmak Nuna*), in Canada's Arctic archipelago, is depicted as roughly the size of Australia. In fact, Australia is over thirty-nine times larger. Greenland occupies the same area as the whole of Africa, even though it would fit inside the continent fourteen times.

The particular pattern of the great land masses and their relationship to each other became hardwired into the global consciousness over the centuries. North is at the top and south is at the bottom. The equator is often drawn two-thirds of the way down the map because, presumably, when Gerardus was flogging his maps around Europe, there was less interest in the southern hemisphere, much of it unexplored or regarded as uncivilised.

In December 1972, an unauthorised photograph helped change our relationship with the planet for ever. I remember seeing it on the front page of almost every paper in the newsagent. They called it the 'blue marble' (AS17-148-22727).

The crew of Apollo 17 were just over five hours from their launch in Florida when one member grabbed a 70mm Hasselblad camera and snapped the view of Africa and Arabia sliding past the window, quite outside the strict schedule and flight plan set down by NASA control. It became one of the most reproduced images ever. As a boy, I had a copy pinned to my bedroom wall.

It wasn't the first picture of the whole sphere of Planet Earth, but there is something about this perfect circle of colour set in the total blackness of deep space that simultaneously suggests vitality and vulnerability. The blue of the oceans, the brown and green of the land in Africa and Arabia, the white of the swirling clouds – it is an image that has caused millions to pause in wonder. It was our little lump of rock, our Ithaca, our island in infinity.

When technicians at the photographic laboratory inside Building 8 of the Johnson Space Center found the 'blue marble' image on one of Apollo 17's Hasselblad film rolls, they assumed the picture was inverted. The astronaut would have been weightless and the spacecraft rotating, they realised, so it was understandable that the world was the wrong way up. When it was published, countless newspapers and magazines rectified the mistake. North was moved to the top. Africa was now correct.

This adjustment, however, was later produced as evidence of how European thinking had become accepted as the prism through which our planet should be viewed. It wasn't always the case. In medieval Europe, many maps were drawn with east at the top. During the same period, Arab cartographers tended to put south where north would be now, as the early Egyptians and the Chinese had done for centuries before them. It was the abiding influence of Ptolemy, maintained by Gerardus Mercator, that probably confirmed the position of north at the top, a global convention that would have a sustained psychological impact on how people see their place in the world.

There is plenty of academic research suggesting a north–south bias, a general sense that the northern hemisphere is superior to the southern – geographically, economically and culturally. In 1979, the Australian artist Stuart McArthur produced an alternative perspective he called 'McArthur's Universal Corrective Map of the World'. South is at the top and Australia is placed in the position one might conventionally find Europe, no longer 'down under' but dead centre.

I look at Stuart's map and notice how uncomfortable it makes me. It feels wrong. I struggle to locate the United Kingdom, which appears tucked away in the bottom right-hand corner, and find myself focusing instead on the countless islands of the Pacific. I become aware of how profoundly we are affected by the traditional western and continental view of our fragile planet. The Mercator interpretation of the earth focuses on the great landmasses of the northern hemisphere, magnifying and distorting their importance. Islands, meanwhile, tend to become lost in the seas which serve to define the continents, relegated to the margins.

In 1998, the Argentinian Post Office (*Correo Argentino*) produced a $0.75 stamp that featured the 'blue marble' image of the world beneath a picture of a human foetus, with the legend *Cuidemos Nuestro Ozono* (Save Our Ozone Layer).

Quite deliberately, the photograph of the planet was reproduced as it was originally shot, with Antarctica at the top.

In August 2018, Royal Mail issued a set of six commemorative stamps to mark the 250th anniversary of Captain James Cook setting sail from Plymouth aboard HMS *Endeavour*, the first of his famous voyages to the South Pacific. The designs were mostly as one might expect: classical portraits of those involved, botanical illustrations, views and maps related to the expedition. But the second-class stamp in the series was odd. It featured a simple pencil drawing described as 'Chief Mourner of Tahiti and a scene with a canoe by the artist of the Chief Mourner (Tupaia)'.

The explanation was cryptic and the image strange, but that final bracketed name 'Tupaia' referred to a quite extraordinary individual: go-between, guide, translator, warrior, diplomat, philosopher, priest. To some, he was close to a god. Who, you might be forgiven for asking, was this mysterious Tupaia?

To answer that question, it is best to begin by seeing the world from a different angle. Take a globe and turn it until the remote South Pacific island of Ra'iātea is placed dead centre. Planet Earth looks very different from this perspective, with the great continents relegated to the margins, the coast of the Americas only just visible on the far eastern edge, Antarctica lurking icily in the far south and a glimpse of Australia to the south-west. Almost the entire sphere appears filled with the blue of the ocean. But take a moment to examine further. The vast sea is not empty but stippled and flecked by islands. Tens of thousands of islands. This is the planet of the islander.

With Ra'iātea bang in the middle of this oceanic circle, let us dive down to the island itself. Here, in around 1725, a child was born who would go on to command a special place in the story of islands and, arguably, in British history. His name was Tupaia.

Groomed for leadership at the island's school of the high priests (*fare-'ai-ra'a-'upu*), Tupaia had a remarkable talent for retaining and recalling facts, able to memorise huge amounts of information about each of the islands scattered in the seas around them: their geography, history, mythology, population, government, harbours, reefs, flora and fauna. Tupaia became equally expert on the stars and the planets, the heavenly arc of the southern sky that helped guide the sea-canoes from one island port to another. But his most extraordinary gift was being able to process all this data in his head somehow to produce a detailed mind map of the universe as it appeared to the people of Ra'iātea, with their island at the very centre.

When Captain James Cook arrived at the largest of what he named the Society Islands (Tahiti) aboard HMS *Endeavour* in April 1769, he was the third European explorer to cross the beach in as many years. British naval officer Samuel Wallis had been there in 1767, shot a number of the local tribesmen, infected the women with venereal disease and christened his conquest King George's Island in honour of his distant monarch. The following year, French Admiral Louis de Bougainville arrived, his ship full of scurvy and mutinous sailors, happy to exchange axes and knives for food and lodging, before renaming the territory the New Cythera (*La Nouvelle Cythère*), after the Greek island where the goddess Venus (Aphrodite) is said to have arisen from the foaming water of its shoreline. A little over a year later, James arrived, also with Venus on his mind. His secret orders were to observe the transit of the planet Venus across the sun, an event the Astronomer Royal had wanted measured and logged from the middle of the South Pacific in order to calculate the precise distances between the planets, thus unlocking the mysteries of the solar system.

It is hard to imagine how these three arrivals in three years, the first ever from continental Europe, must have affected the islanders of Tahiti as they stood semi-naked on the shoreline. Philosophically, culturally, socially and spiritually, the aliens striding purposefully across the beach of Matavai Bay, with their breeches and buttons and braid, were visitors from another world, terrifying and fascinating in equal measure. For all those on that island boundary, ancient certainties were about to be challenged.

The Europeans had two ambitions: territory and knowledge. The kings of both England and France intended to plant their flags in the previously unmapped lands of the southern hemisphere, including, they hoped, the widely hypothecated sixth continent *Terra Australis Incognita* (Unmapped Southern Land). It was a mark of their continental prejudice that European leaders assumed there must be a large land mass somewhere in the southern hemisphere to balance the great continents of the north. Locating and claiming it was a strategic geopolitical aim to be backed up with cannons and muskets as necessary. But in Europe's Age of Enlightenment, these expeditions were also about science and academic study, particularly in the case of James Cook and his team aboard the *Endeavour*. Supported by the Royal Society, James had invited a team of scientists to join him on the voyage – an astronomer, three botanists and two illustrators who would assist in documenting the discoveries. Their work, it was anticipated, would provide vital intelligence to further Britain's power and prestige.

Leading the scientific party was botanist Joseph Banks, still only in his mid-twenties but with a reputation for the rigorous pursuit of knowledge across the natural sciences. From rocks and plants to people and the stars, he was fascinated by it all. Joseph had invited kindred spirits to join him on the *Endeavour*, including his good friend Daniel Solander, a pupil of Carl Linnaeus (*Carolus Linnæus*), the revered Swedish botanist who had recently formalised a two-word system for the classification and naming of all living things. Daniel was one of the so-called Apostles of Linnaeus, students who had been given the blessing of 'the father of taxonomy' to use his approach on scientific expeditions around the globe. Joseph had readily agreed to use Linnaean classification, giving Latin names to all he might discover in the South Pacific.

With the ship docked at Tahiti for a few months to allow the construction of a fort and observatory for the transit of Venus, there was an opportunity for Joseph to gather information about the island's plants and animals, as well as the customs and language of the local people. He knew his task would be much simpler if he could tap into the native wisdom by finding an island guide who could be persuaded to help translate both linguistically and culturally. Step forward Tupaia.

Tupaia had moved to Tahiti having fled his home on Ra'iātea because of a war between the islands, arriving in time to witness the arrival of the European voyagers. He had observed them carefully, quickly picking up something of their languages and analysing their strange ways. As a high-ranking priest of the Arioi cult, he would have been an imposing figure, tattooed from his thighs to his heels, marking him out as an esteemed 'black leg' (*'avae parae'*) and wearing the grand regalia of an aristocrat and spiritual leader.

A friendship (*taio*) blossomed as the islander introduced the botanist to the natural wonders of Tahiti, including the sacred white flower called *tiare mā'ohi*, a fragrant bloom said to have been a gift from the god Ātea, encouraging humankind to abandon prideful ways and find love and harmony in all they did.

The tribal names for places and plants on the islands of the South Pacific often contained hidden layers of meaning and stories, a poetry intrinsic to an oral culture, where the sounds and rhythms of the spoken word had far greater significance than in societies with a strong written tradition.

Joseph decided to rename *tiare mā'ohi* as the Tahitian gardenia (*Gardenia taitensis*), honouring the Scottish botanical pioneer and Linnaean enthusiast Alexander Garden.

As the crew of the *Endeavour* prepared for the disc of Venus to journey across the face of the sun, Tupaia charmed the ship's chief scientist with his stories and his knowledge, amusing him with sketches of island life that would later be incorporated into the design of a commemorative postage stamp.

In the same way that the Europeans had territorial as well as scientific ambitions, Tupaia's interest in the expedition included a strategic element. The centre of his cosmos was his island, Ra'iātea, hallowed territory upon which a shadow had fallen in the form of an occupying army, warriors from the Tahitian island we now call Bora Bora (*Pora pora mai te pora*). He hoped that the Europeans with their firepower might be persuaded to help him oust the invaders. Joseph was captivated by Tupaia and even agreed to let the priest inscribe his body with an

Arioi tattoo, a permanent mark of Pacific island culture upon western flesh.

'The captain refuses to take him on his own account,' Joseph wrote in his log, resolving to pay for Tupaia's stay on board the *Endeavour* himself. 'I do not know why I may not keep him as a curiosity, as well as some of my neighbours do lions and tigers,' he wrote. 'The amusement I shall have in his future conversation and the benefit he will be to this ship will fully repay me.'

In July 1769, HMS *Endeavour* left Tahiti with Tupaia expertly piloting the vessel through a maze of reefs and atolls without chart or instrument. He guided them northwest to his own island of Ra'iātea among the volcanic outcrops of what would later be called the Leeward Islands. The captain's journal explains how the priest conducted a peace ceremony involving handkerchiefs, beads and feathers, before James 'hoisted an English jack, and took possession of the Island and those adjacent in the name of His Britannick Majesty, calling them by the same names as the natives do'. Disappointingly for Tupaia, James could not be persuaded to engage the forces occupying his homeland. The explorer was keen to keep exploring and so Tupaia advised him to take a bearing south across open ocean for almost 400 miles until, precisely as predicted, the outline of Rurutu, one of the Austral Islands, loomed up from the horizon. James was so impressed by Tupaia's navigational skills that he invited him to the ship's drawing table for a session in map-making.

A large sheet of paper was laid out, upon which, it was hoped, the extraordinary priest might help create a detailed and accurate chart of the Pacific Ocean. Tupaia looked at the blank page and the expectant eyes of those gathered around, realising the fundamental problem he would have in reconciling the mental map of the Polynesian island dwellers with the formal cartographic traditions of Europe.

For those who lived on continental land masses, the relationship between places was defined by the physical and measurable spaces between them, literally the number of paces it would take to walk from one spot to another. Maps attempted to represent the arrangement as though the surface of the earth was flat and one was looking down from above, with the landscape fixed. The distance from Rome to Paris was precisely the same as the distance from Paris to Rome.

This was not how the islanders of the South Pacific viewed the world. Distance was not a factor of space but of time, measured by the hours and days an outrigger canoe would take to travel between islands. The winds and currents of the ocean would mean that two ports could be much closer going in one direction than another. This geography was not solid like the land but fluid like the ocean, shifting with the seasons and the weather. The fixed point was the cockpit of the canoe (*pahi*) around which the universe swirled. The direction of the voyaging routes was set by the stars, the currents, the clouds and even the birds and fishes. Instead of a map, they relied on metronomic chants, ancient oral mnemonics that jogged the memories of navigators as the rhythm of their oars pulled them towards their destination.

Tupaia tried to work out how to reconcile these two distinct visualisations of the cosmos and came up with an ingenious solution. He asked that the word '*avatea*' ('noon') be written at the centre of the map. This point, he explained, would mark north when the sun was at its height from wherever on the chart you were. To navigate to any of the islands, you must first take a bearing to *avatea* and then a second to your destination island along defined voyaging routes as indicated by the map. The angle between the two would set your direction.

It would take centuries before the brilliance of Tupaia's compromise was properly understood, but the commanders on the *Endeavour* were enthralled as they watched the high priest place dozens of remote islands on the empty plan: from Rotuma on the left margin to Rapa Nui (Easter Island) on the right, from Hawai'i at the top to Rapa Iti (Bass Islands) at the foot, a speckled seascape of named island communities extending across 20,000 square miles of ocean. If there was disappointment amid the excitement of the British observers, it was that Tupaia drew only islands and not the great continent they were under orders to locate.

There is something about an index, the wonderful logic of an A to Z, neatly categorising and logging and sorting according to clear rules and protocols, that is deeply satisfying. I flick to the back of my large atlas and run my finger down the list of places around the globe.

Aach Germany
Aadorf Switzerland
Aalborg Denmark
Aaley Lebanon
Aarah Maldives
Aasiaat Greenland

The varied influences of language and culture, power and politics can be divined from the inventory. It tells a story in alphabetical order.

George Town *Tasmania* Australia
George Town *Great Exuma* Bahamas
George Town *Grand Cayman* Cayman Islands
Georgetown Gambia
Georgetown Guyana
George Town Malaysia
Georgetown *Connecticut* USA

Scanning the pages, I realise it is possible to track the voyages of colonising European powers, as they applied an imperial stamp to their conquests, honouring their monarch and planting their flag.

Victoria Grenada

Victoria *Gozo* Malta

Victoria *Mahé* Seychelles

Victoria, Mount *Viti Levu* Fiji

Victoria, Mount *South Island* New Zealand

Victoria, Mount Papua New Guinea

Colonial nomenclature was slapped like military whitewash on almost every community the explorers encountered, obliterating the native names of places and features, plants and animals. At an informal level, European sailors would nickname the reefs and bays they navigated, the wildlife they saw. But there was a more deliberate naming process that sought to embed the authority of the Crown into the geography and biology of conquered land.

I look at Pangaea as she lies in her meditative trance upon the desk, her weight holding down the right-hand side of my map, countless islands of the South Pacific scattered in front of her. It is noticeable how labels attached to the strings of territories arrayed across the ocean are derived from the family names of eighteenth-century officers and crew of British ships. Robert Pitcairn, John Marshall, Thomas Gilbert and, of course, James Cook are awarded the honour of their own archipelago. The native islander names, resonant with the rhythms and cadences of an oral tradition, often sit beneath the European appellation.

Among the Cook Islands, previously called the Hervey Islands to honour British Admiral Augustus Hervey, is an atoll the Europeans marked on their maps as Armstrong Island. It already had a name, Rarotonga (The Down South Island), given to it by the Samoan nobleman and voyager Karika hundreds of years earlier. On the western coast of Rarotonga is a settlement called Te Rerenga Vaerua, a phrase singing with mystical meaning, describing a spot where spirits leapt from the rocky reef on the shoreline to join the path of the setting sun to the underworld.

In 2019, a 'name change committee' was assembled, searching for a new name that would reflect the history and culture of the Cook Islands' indigenous Māori inhabitants, the suggestions to be put to a public vote. The idea is not an original one. A referendum in 1994 saw the proposal for a new name rejected, but in recent times the tribal chiefs from the fifteen islands have been singing more harmoniously on the need for the territory to reflect its story before the Europeans arrived.

It is not just the geography of islands that has been scrubbed of its native markings. The indigenous plants and birds, insects and animals that make up unique island ecosystems have been categorised using rules contrived by Europeans. The binomial system of classification devised by Carl Linnaeus was intended to bring scientific logic and structure to the natural world, but its Latin constructions were also often used to honour white men with a classical education rather than respecting the folklore and poetry of native cultures. I realise, that for all its rationality and reason (so appealing to my sense of order), Linnaean taxonomy misses the essential truths secreted in the myths and the magic of island folklore.

In the map room on board HMS *Endeavour*, the clash between continental and island worldviews became a straight argument over direction of travel. Tupaia promised to introduce James and his colleagues to the wonders of many previously unmapped islands if they sailed west. The captain, however, had orders from his European paymasters to locate *Terra Australis Incognita*, the vast landmass which it was supposed must lie to the south. In the end, the *Endeavour* took the latter course and a miserable Tupaia withdrew to his room. It was a bitter humiliation for the Polynesian nobleman, his navigation and translation skills no longer required as continental priorities held sway.

On 6 October 1769, the *Endeavour*'s lookout, high on the main mast, reported land had been sighted, territory close to an area previously mapped by Abel Tasman more than a century earlier. The Dutch explorer had postulated that what he claimed as *Staten Landt* (Land of the Dutch States-General) might be the western coast of a huge continent stretching across to South America, but he never actually set a European foot upon the shore, a fatal skirmish with local Māoris in the coastal waters convincing him to move on without making landfall. Abel sailed away from *Aotearoa* (New Zealand), having named the site of his encounter *Moordenaars Baij* (Murderers' Bay).

When the *Endeavour* dropped anchor in a bay a few hundred miles further north, Tupaia remained in his cabin. James, Joseph and Daniel decided to take a small boat and go ashore, unaware how they would miss the priest's island wisdom. Before nightfall, one of their guards had shot dead a man who appeared to be threatening them with a spear. The deceased was Te Maro, a prominent leader of the Ngāti Oneone tribe who, it is now believed, was offering a ceremonial challenge to the uninvited strangers who had arrived from their floating island. The explorers returned to their ship, listening to the desperate cries of the Māoris gathering around the body on the beach.

The following morning, James resolved to try to open up friendly relations with the locals and sought Tupaia's help. The priest was among the party of musket-bearing Europeans staring across a river at an angry and hostile band of Māori tribesmen shortly afterwards.

Tupaia's genius was as a cultural interlocutor, blessed with the unique ability to bridge the river of understanding that lay between the two groups that morning. To the amazement of James and his men, he was able to communicate with the natives on the far bank, skilfully navigating the current of shared language that had flowed along the voyaging routes of islanders over centuries. The priest heard the echoes of tongues and traditions from his own island thousands of miles across the ocean, felt the rhythm of the wider archipelago, recognised the secret meanings that underpinned the motivation and the behaviour of the people over the water. Tupaia used his talent as translator and diplomat to calm the situation and, while warning James to remain wary, encouraged one of the Māoris to strip off and swim to a small islet in the river. A short time later, the tribesman stood upon a rock named Te Toka-a-Taiau, a spiritual gathering place and boundary stone, beckoning James to join him. The captain made his way and, in an act dripping with symbolism, the Māori greeted the European in the time-honoured tribal fashion, with a *hongi* (nose-touching greeting). At that moment, two breaths and two worlds were joined into one.

The spirit of peace and goodwill was to be short lived, however. Before nightfall, five more Māoris were dead following what may have been another 'cultural misunderstanding', this time over the exchange of weapons. Tupaia's role as mediator became critically important. In the next few days, he neutralised a potentially deadly encounter with Māori war canoes, calmed tempers after yet more shootings, conversed at length with local leaders and slowly won their trust, respect and, ultimately, admiration. To them, Tupaia represented a thread connecting the Māori to their ancestors across the ocean, the living embodiment of an island heritage from which they had been severed for hundreds of years. The first canoes had arrived in New Zealand from eastern Polynesia in the fourteenth century.

He was treated as an honoured guest, draped in ceremonial cloaks and entrusted with sacred treasures. Huge crowds came to hear him preach. To this day, Tupaia is feted by the indigenous Polynesian people of New Zealand as a hero.

James Cook did not mention the *hongi* in his journal. Instead, he focused on the 'unfortunate and inhospitable' place they had landed. He named it Poverty Bay, 'as it did not afford us a single article that we wanted'. There was more than a hint of retribution in his decision to replace the native name *Te Oneroa* (Long Sand) with a negative English label, an act of deliberate cultural vandalism.

In the past few years, an online atlas was produced restoring customary Māori names to numerous places across what was increasingly called 'Aotearoa New Zealand'. In 2019, as the country marked the 250th anniversary of James Cook's landing, a memorial to Te Maro was unveiled close to where he was shot, and after months of negotiations with local tribes, the British High

Commissioner expressed public 'regret' (but no apology) for the killing of Māori people after the arrival of HMS *Endeavour*.

Tupaia died after an illness in 1770, most probably dysentery or malaria, contracted aboard ship. His voyage was over.

CHAPTER 12

ISOLATION

Barriers and Bridges

I stand on the beach beside a metal man. Together we stare out to sea. The metal man has no eyes. But he stares as I stare. We are not alone.

A BBC cameraman is filming us standing on the wet sand, me and 100 life-size, cast-iron figures spread out along a chilly and blustery Crosby Beach on England's north-west coast. A drone buzzes overhead, getting the view from above as seagulls screech. Antony Gormley's unsettling sculpture *Another Place* has drawn me to this stretch of foreshore for a report on identity in Britain. I want to consider how living on an island affects the way we think of ourselves. In my overcoat and carrying a black briefcase, I look like a fish out of water as the tide washes over my unsuitable shoes. I notice the metal man is dotted with barnacles, a pale stain on his rusty face resembles the track of a tear.

The Danish word for 'island' is a single letter: Ø. The iconicity pleases me. The circle of the O is the shoreline, the boundary between itself and everything else. But the slash suggests the line of insularity has been broken, cancelling out the perfection of the circle. An island, the one-letter word graphically suggests, is isolated and not isolated at the same time.

Like the sand and sea that swirl at my feet on Crosby Beach, Ø is an amalgamation of distinct elements, the blended sounds of an O and an E. I find myself thinking of the relationship between Pangaea and Panthalassa, a circle of land (O) merging with the ocean (E), symbolised by the trident of the sea-god Poseidon (Neptune). Ø is a gliding vowel, two articulations combined, a diphthong (double tone).

The word 'island' itself also hints at a fusion of liquid and solid with its echoes of the Old English *éa* (water) and *lend* (land), a noun with roots leading back to Neolithic times. The contradictory status of an island is present when simply uttering the word.

The island of Dejima was the invention of a worried shōgun. In 1634, Tokugawa Iemitsu, the most powerful man in Japan, ordered a team of Nagasaki workers to smash through a small peninsula that protruded into the city harbour, creating a fan-shaped artificial islet just off the country's south-western coast. Once the fingernail of land had been snipped off the isthmus, a narrow bridge was constructed, heavily guarded at both ends. A squad of gatekeepers and nightwatchmen controlled by a fearsome supervisor (*otona* – 乙名) was commanded to secure the crossing at all times. Almost no one was allowed to travel to Dejima from the city of Nagasaki on the main island of Kyushu, and almost no one was permitted to travel to Nagasaki from the shōgun's isolated creation in the bay. Those who attempted to defy the rules faced execution: methods included being boiled alive, burned at the stake, crucifixion, decapitation or being sawn in half.

Japan had long defined itself by '*shimaguni konjo*' ('island country mentality'), a phrase which translated culturally as a shared commitment to protect and sustain the distinctive characteristics of the nation.

In the seventeenth century, upon the four main islands and numerous smaller ones along the archipelago, there were growing fears that western influence was threatening the unique qualities of Japanese life. Their coastal waters were full of foreign trading ships and pirates, while their towns and cities were bustling with exotic goods and ideas. Jesuit missionaries were actively trying to convert the country to Christianity, sometimes supported by pious European merchants who made loyalty to the church a condition of trade. Several influential feudal lords (*daimyōs*) on the largest island of Kyūshū converted to Catholicism, a move interpreted as a direct challenge to the power structures and traditions of the state.

By the time Iemitsu became shōgun in 1623, military leaders had deported many missionaries and crucified prominent proselytisers and converts, but public anxiety at the unsettling changes introduced by globalisation had reached new levels.

Iemitsu, fearful that his own authority was threatened from within and without, decided to respond with a ruthless display of power. He ordered virtually all Europeans to be expelled from Japan. Extensive restrictions on imports and exports were imposed, the flow of information in and out of the country was all but banned and, to make his intentions quite clear, when two Portuguese sailors pleaded with him for trade to continue, Iemitsu had them executed.

An army of spies and local informants were recruited to expose any Catholics still operating. The entire population of Japan was forced to register as parishioners of Buddhist temples, with suspected Christians ordered to tread on pictures of Jesus or the Madonna. If they showed the slightest reluctance, they were killed.

It wasn't just tight controls on anything foreign crossing the border into Japan. Almost no Japanese citizens were allowed to leave the islands. Those nationals who had been travelling abroad when the decree was passed were unable to return. Any violation was also punishable by death. The policy would later be called the edict of *sakoku* (country locked in chains – 鎖国), implying that

Japan had become entirely isolated, that the islands were closed.

But no island is ever totally detached.

The old Japanese saying '*shiranu ga hotoke*' literally translates as 'not knowing is Buddha'. Ignorance is bliss. The proverb urges the faithful to give up on the pursuit of understanding in order that heart and mind can be undisturbed, at peace like Buddha himself.

But Iemitsu recognised that knowledge was power and, as he pulled the shutters down and raised the drawbridge on the outside world, the shōgun must have feared blindsiding himself to potential threats and missing out on opportunities and pleasures. He needed a back channel, a discreet portal to allow some contact with the west, a relationship on his own terms. His answer was to create the island of Dejima.

For more than 200 years, the narrow bridge between the island and Nagasaki city would be the only official link between Japan and Europe. Dejima comprised little more than a small village of wooden houses, offices, storerooms, warehouses and wharves, initially populated by Portuguese merchants and staff. However, five years after they had taken residence, Iemitsu suspected their involvement in an uprising by 40,000 Christian peasants and beheaded sixty-one of them, deporting the rest. The Dutch East India Company (*Vereenigde Oostindische Compagnie* – VOC), which had shrewdly provided gunpowder and cannons to help the shōgun crush the rebellion, was moved from its regional headquarters on the island of Hirado to replace them.

It was a strange existence for the Europeans on Dejima. They had virtually no contact with their Japanese hosts apart from workmen, cooks and officials deemed trustworthy enough to cross the bridge. After a few years, handpicked courtesans from the Maruyama teahouses were sent to the island to attend to the community's needs, their visits earning them the nickname *orandayuki* (those going to Holland) and an opportunity to smuggle a little sugar and silk back over the bridge.

Despite the isolation, the operation at Dejima proved very profitable for the VOC and the employees were instructed to endure hardship and even humiliation to keep the shōgun sweet. Every year, company representatives were required to pay homage at his castle in Edo (present-day Tokyo), taking the most extravagant and extraordinary gifts.

Government officials would describe in advance what their master desired, often the latest innovations in western science: astrolabes, telescopes, spectacles, medical instruments and globes. The shōgun would pre-order books, medicines and a menagerie of exotic creatures that had to be unloaded on Dejima, taken across the little bridge and transported more than 700 miles overland to the capital. Camels, a civet cat, cassowaries, cockatoos, buffaloes, racehorses and one of the last surviving dodos (*Raphus cucullatus*) were imported as tributes to the court.

This bizarre procession of animals and birds, commodities and traders, winding its way across the islands of Japan symbolised the intoxicating allure of globalisation. To those with money and power, western capitalism promised the accolade of specialness. Every rare and luxurious item set the shōgun apart from the masses, a mark of distinction and superiority in a sea of oblivious subservience. Isolation was fine for those outside the palace walls, but the elite needed to stay connected.

When Iemitsu's son Tsunayoshi became shōgun, he would demand the VOC executives perform Dutch dances and songs when they came to deliver their gifts at the castle in Edo. However mortified the businessmen may have felt as they hopped up and down uncomfortably, the indignity was worth it when they scanned the corporate accounts.

Each of the 100 expressionless iron men surveying the Irish Sea is cast from a mould of Antony Gormley's own body, a literal depiction of 'oneself' installed on the shoreline to reflect upon 'another place', or so the sculptor's title suggests. Identity is a relational phenomenon, how 'self' relates to 'other'. Here, on the beach at Crosby, I want the cameras to capture the territorial version of that association: an island clearly delineated from the outside world by its physical geography, the encircling sea creating what appears to be an emphatic boundary between 'us' and 'them'.

I follow the metal man's blind gaze to the horizon and a boat taking the familiar journey from Dublin to Liverpool. Ferries have ploughed the same sea lane back and forth over centuries, transporting people and their baggage across the water, shaping the character of two great cities on their respective islands. It can sometimes be a perilous journey, the thousands of shipwrecks littering the seabed a memorial to the perils of leaving home ground, dangers that did not cease even when a ship was safely tied to the harbour.

The arrival of coronavirus in 2020 was a reminder for people living on the banks of the River Mersey of Black '47, the 'calamitous' typhus outbreak of 1847 in which 'Irish fever' was blamed for turning Liverpool into the city of the plague. In the first six months of that year, 300,000 destitute refugees escaping famine and disease crowded onto steamers from Dublin and crossed the water from one British isle to another. Those arrivals who appeared sick were housed in isolation sheds and warehouses close to the quayside or held on quarantine ships offshore. By June, with tens of thousands of people dead and hospitals overwhelmed, the British Parliament passed a law allowing the authorities in Liverpool to deport 'Irish paupers', UK citizens whose status had changed from 'us' to 'them'.

The grey outline of the ship on the horizon off Crosby Beach is a symbol with two meanings, two sounds. A diphthong. It represents the potential threat to the health and integrity of the island community but also signifies the benefits of new blood and fresh thinking. The boundary where water meets land is both a barrier and a bridge, a membrane that can be open or closed.

A number of western powers became almost tormented by their inability to access the little island of Dejima and the lucrative Japanese market that lay beyond. Over the next two centuries, Portuguese, American, Russian, French and British ships all tried and failed to engage with Japan's rulers, hoping to break the Dutch monopoly. Diplomacy and aggression were met with the same implacable response.

There was also a sense that, through its policy of isolation, Japan was thumbing its island nose at the values of the great powers in the Christian west. This idea had particular traction in the United States, a nation born and shaped from a belief in continentalism, a certainty in its God-given manifest destiny to control the whole landmass of North America from sea to shining sea. It was a conviction that spilled beyond its Pacific coast and across the ocean, translating into a moral duty to impose what it regarded as its civilised and superior worldview upon the backward islands of the Orient.

In 1851, the US President Millard Fillmore signed a letter addressed to Emperor Kōmei of Japan, reminding him how America now extended 'from sea to sea', incorporating the 'great countries' of Oregon and California with their riches of gold, silver and precious stones. He explained how American steamers could 'reach the shores of your happy land in less than twenty days', passing close to Japanese waters on their trade route to China, and how the US government would 'ask and expect' the emperor to protect American property, to offer a harbour for their ships, to allow the purchase of coal and to sanction trade. 'Our object is friendly commercial intercourse, and nothing more,' the President lied. The letter was, in reality, a barely veiled threat to be backed up with continental military might.

On 8 July 1853, two American warships, *Susquehanna* and *Mississippi*, steamed past the Japanese border posts and into Edo Bay, black smoke billowing from their funnels as the paddles churned up the territorial waters of the Uraga Channel. Bristling with the latest in iron-clad cannons, the navy frigates were supported by two sloop-of-wars, *Plymouth* and *Saratoga*, an intimidating display of force. In charge of the small fleet of what the Japanese called 'black ships' was Commodore Matthew Perry, his orders to end two centuries of Japanese isolation with the use of gunboat diplomacy if necessary. His first responsibility, however, was to deliver the letter from President Fillmore.

The US warships were quickly surrounded by Japanese guard boats, one displaying a sign, written in

French, demanding the Americans leave immediately. Matthew was unimpressed. He had realised that the islands' military defences were no match for the weaponry aboard his vessels, an arsenal that included formidable Paixhans guns which fired the latest in explosive shells. The commodore's plan, however, was to bully rather than bombard. He instructed the cannons aboard his ships to open fire with blanks, ludicrously pretending the orchestrated cacophony was part of a belated Independence Day celebration. When Japanese government representatives attempted to engage with the Americans noisily trespassing in their harbour, Matthew refused to leave his cabin because they were not of sufficient seniority to receive the presidential missive. Instead, he instructed a crewman to hand the officials a white flag signifying non-hostility, while also delivering a note warning that should the Japanese put up a fight, they would be vanquished.

East and west, ancient kingdom and modern republic, mysterious island nation and ostentatious continental superpower, two opposing cultures faced each other across the grey waters of Edo Bay wondering who would blink first, the isolationist or the expansionist.

On 14 July, Matthew straightened his commodore's uniform as the *Susquehanna* fired a thirteen-gun salute. He stepped onto Japanese soil as the band struck up 'Hail Columbia', saluting the ranks of 200 US sailors and marines who presented arms as the naval commander marched up the beach at Kurihama. Matthew walked forward and, with great ceremony and reverence, presented the letter from President Fillmore to the emperor's envoys. Although the Japanese had decided that simply accepting a letter did not violate their national sovereignty, the pomp and patriotism the Americans had draped over the whole event told a different story. As the band played on the beach, it was clear that Japan's wall of isolation had been breached.

The following year, Matthew returned with greater menace, more ships, more cannons and more men. This time the orchestra played 'The Star-Spangled Banner' as the commodore walked up the beach and into the purpose-built hall for the signing of a treaty that agreed all the demands in the President's letter. As the signatures dried, presents were exchanged. The Americans gave a hundred gallons of American whiskey, clocks and books about the United States. The Japanese responded with porcelain goblets, gold-lacquered furniture, bronze ornaments and a collection of seashells, a special gift for Matthew.

In July 1953, exactly 100 years after President Fillmore's threatening letter was handed over, the US Post Office issued a 5c commemorative stamp (green) to celebrate the successful mail delivery. It depicted Matthew's 'black ships' at night in Edo Bay with Mount Fuji in the background, a portrait of the commodore in the top right corner and the legend 'Centennial of Opening of Japan' running along the bottom edge.

In February 2017, with Japanese and Dutch royal families in attendance and great ceremony, a narrow footbridge was opened linking Dejima to Nagasaki city. After the arrival of the American ships, the little island lost its purpose and was gobbled up by the port, almost forgotten as global trade expanded. The original bridge was dismantled, the outline of the island buried beneath urban infrastructure. But the area has since been designated a Japanese national heritage site, a shrine to the tensions and contradictions of isolation and islandness.

CHAPTER 13

LAND

Custodians of Eden

I am leafing through some old family photographs discovered in a box in the attic. One is a picture of my father in 1959, the year I was born. He is in the garden of our home in Balmoral Drive on the outskirts of Glasgow, a pipe in his mouth and glasses on his nose, chatting to a neighbour over a low privet (*Ligustrum ovalifolium*) hedge. I can feel Pangaea watching me from her couch as I study the image.

It is a conversation clearly conducted on equal terms: two men, each in their own domain but able to engage contentedly with the other. The hedge, neatly clipped, marks the territorial boundary, the wall around their respective castles. Balmoral Drive is a suburban archipelago of private islands where neighbours may look across the manicured shorelines, admire the flowers (*Chrysanthemum morifolium*), discuss the weather, ask after the baby, share a joke, with sovereignty maintained throughout.

It is a classically British scene, played out on countless residential streets to this day. Our blessed plots reflect something profound about the character of this sceptred isle. Lawns with perfectly straight stripes and trimmed edges, carefully weeded beds with plants staked and labelled, walls and hedges, fences and trellises, constructed to contain the exuberance of nature. Our obsession with gardening and gardens reflects a need to find order in chaos, to classify and catalogue, to take control of our land.

The gardener could not believe what he was being told, shock etched across his jet-black features. The two historians seated with him having lunch were adamant. The land he owned on his island across the water, the sacred island territory passed down to him over countless generations and thousands of years, did not belong to him. It was the property of the Crown.

Koiki (Eddie Mabo) looked in disbelief at the government map. There was the island of Mer (Murray Island) at the eastern end of the Torres Strait archipelago, some 115 miles north of the Australian mainland. There was his village, a few houses draped along the shoreline. There was his beloved garden, extending up the hillside. There was the official description: ABORIGINAL RESERVE.

'It was as though I had punched him in the face,'

Professor Henry Reynolds later recalled. 'He looked angry, aghast, incredulous. How could the white fellas question something so obvious as his ownership of his land?'

Koiki left the lunch consumed by fury. It was an anger that would nourish an undying resolve to right a wrong, to take on the white fellas who had stolen his birthright. From that meal in a study at James Cook University in Northern Queensland in the 1970s emerged an islander who would dedicate the rest of his life to challenging and defeating continental might.

I find another picture in the box, a photograph taken in the early 1960s. This is of my grandfather John, the one who wrote books about British postage stamp design. He is standing on the precisely mown lawn of his house in Scotland, staring straight at the camera wearing a crumpled tweed jacket, one button hastily done up and with pockets bulging. John carries the look of a man who has been interrupted. Perhaps he had been working on another of his books, a volume dedicated to cacti and succulents. I have only a hazy memory of him, sitting in his study, far too busy to play with small boys.

When not at his imposing desk, John was to be found in his greenhouse, where he took enormous pride in the cultivation and propagation of exotic plants. To him, the joy of a shrubbery was similar to the thrill he found in a stamp album. A garden was a collection, in the way that Kew Gardens is home to the national botanical collections, with specimens gathered from around the world, identified and categorised.

Britain's national fascination with gardening is, in part, a legacy of the botanists who rode on the coattails of the expanding empire to hunt for new and rare plants. The inspiration was, of course, Joseph Banks aboard Captain Cook's HMS *Endeavour*. Joseph sent back more than a thousand species never previously seen in Europe, many of them unique flora found only on certain islands. He sent his botanist protégés off to explore many island bionetworks: Francis Masson was despatched to Madeira, the Canary Islands, the Azores and the islands of the West Indies; David Nelson was aboard the ill-fated HMS *Bounty* under instruction to collect breadfruit plants (*Artocarpus altilis*) from Tahiti; William Hooker was put on a boat to Iceland to bring back specimens, while his son Joseph took botanical island-hopping to the next level, sending rare plants to Kew from dozens of islands he visited on an epic voyage in 1839. He even stopped at the Kerguelen Islands (Desolation Islands) in the Antarctic, described as the most isolated islands in the world, where Joseph identified eighteen flowering plants, thirty-five mosses and liverworts, twenty-five lichens and fifty-one algae, many never previously recorded.

To my grandfather John, I suspect, gardening had associations with exploration and discovery, with a fabulous idea of civilising the world and taming the wilderness.

Exactly 100 years before Koiki's fateful meal at the university, Tudu islanders reluctantly showed Tongatapu Joe the shell beds on Warrior Reef and changed the lives of the people of the Torres Strait for ever.

Captain William Banner, an experienced master pearler and bêche-de-mer trader operating in the Pacific, was away in Sydney on business. He had left his lugger *Bluebell* in Joe's capable hands and the Tongan diver was scouring the seabed in the hope of stumbling across a reef that was home to the prized black teatfish (*Holothuria nobilis*), the sea cucumber or bêche-de-mer regarded as a slippery delicacy in south-east Asian cuisine. Joe also had half an eye open for rare silver-lipped oysters (*Pinctada maxima*), reef-dwellers which yielded the very finest mother-of-pearl, their large flat shells decorated on the inside with a gleaming silver-pink and champagne nacre. The pearl was highly sought after in Europe and America, used for fancy buttons and buckles, combs and cufflinks, revolver handles and furniture decoration. There were fortunes to be made and it was a vicious business.

Joe had set out from the tiny island of Tudu, where William had just established a processing station for bêche-de-mer. Almost two-thirds of the local tribespeople had fled when the *Bluebell* first arrived, the reputation of the master pearlers preceding them. Men, women and children from the island's forty-three families simply boarded canoes and headed out into the ocean rather than encounter the lugger crew. Of those who remained, some were put to work in the factory and others persuaded to reveal the secrets of their fishing grounds. Joe had demanded to know where they obtained the pearl shell he had seen used for jewellery and ornaments. The answer was Warrior Reef.

There in the shallow waters, just below the surface, Joe could see huge numbers of silver-lipped oysters clinging to the coral. It was an astonishing sight, a treasure trove just waiting to be picked up. By the time the captain returned to Tudu, the crew of the *Bluebell* had harvested six tons of oysters, a cargo worth a quarter of a million pounds in today's prices.

The owner of the *Bluebell*, Sydney-based entrepreneur James Merriman, told William to stop diving for sea cucumbers and focus solely on pearl. He sent more of his boats to the area and scooped up fifty tons of silver-lipped oysters within a matter of months. News of the discovery sparked an oyster-rush as master pearlers from across the region were ordered to steer their lugger fleets to the Torres Strait. There was a frantic search for workers to crew the boats and dive onto the reef, a demand that triggered a wave of 'blackbirding' along the archipelago.

Blackbirding has been described as 'Australia's dirty little secret'. For decades after slavery had been abolished by the UK Parliament, unscrupulous colonists kidnapped and enslaved tens of thousands of Pacific islanders to work on plantations and sheep stations, in guano mines and aboard the fleets of luggers looking to exploit the treasures of the southern seas. The captains of blackbirding ships could get five shillings for each able-bodied young man they recruited, seized or tricked onto their vessels and delivered to the Australian mainland, sometimes hanging numbered discs around each islander's neck to make it easier for buyers. Prices on the quayside ranged from £4 to £20 a head, many of the individuals no more than terrified teenage boys whose chances of surviving to adulthood had diminished significantly.

News of Tongatapu Joe's harvest from Warrior Reef prompted the blackbirders to target the nearby islands of the Torres Strait for pearling crews. At the eastern end of the archipelago, Mer Island had a taste of the people traffickers' methods. Raiders seized around 400 men and women, murdering those who resisted. A way of life that had barely changed for millennia was close to being crushed in the continental stampede for a quick profit.

These were desperate days for the islanders of the Torres Strait, who pleaded with their gods to rescue them. As if in an unexpected answer to their prayers, as the sun set on 1 July 1871, a ship named HMS *Surprise* dropped anchor off the coast of *Erub* (Darnley Island), a small community huddled in the north-east of the archipelago. A dinghy was lowered, and two vicars clambered aboard, the Reverends Samuel McFarlane and Archibald Murray of the London Missionary Society (LMS), looking for souls to save. In a dramatic scene now recreated every summer on the beach of Erub, Samuel splashed his way onto the volcanic sand and fell to his knees holding his bible aloft. Fearsome Erubian warriors watched the arrival from the undergrowth, spears at the ready as a tribal elder named Dabad approached the praying priest. The expectation was that he would instantly behead the unfortunate evangelist and keep his skull for the power and strength it would bestow upon him and the tribe.

In that moment upon the shoreline, Dabad had a calculation to make. With the islanders facing a potentially existential crisis from profit-hungry traders, did the supine figure represent a threat or an opportunity? Was this the moment to lower or raise the drawbridge? It was, in a sense, the great challenge of islandness. The audience hidden behind the hedge of vegetation looked on aghast as, contrary to centuries of tribal law, Dabad dropped his weapon and welcomed the priest to the island.

The arrival of the LMS was a cultural turning point for the Torres Strait, regarded by many to this day as 'the coming of the light'. The history of the islanders was cleaved into two; there were the old days when the islands were governed by spirits and ancient magic, a period known as '*bipotaim*' ('the before time'); and there was a new era of western Christian order.

The presence of the missionaries helped bring an end to the ferocious inter-island raids that had terrorised communities along the archipelago, it offered protection from the threat of rapacious pearlers and blackbirders and it provided education and employment. The church even set up its own pearling business designed to promote 'native enterprise'. Despite anxiety that they were being detached from their tradition, most islanders took the plunge of baptism and converted.

The missionary leaders posted to the Torres Strait had largely been recruited from congregations on distant South Pacific islands where obedience to the Bible had an authoritarian character. Flogging, head-shaving and exclusion were common punishments for fornication, quarrelling or failing to observe the Sabbath. The old tribal customs, songs and dances were regarded as blasphemous.

The island of Mer was selected as the location for a regional theological centre, training local men as priests and permanently embedding the strict rules of the mission into island life. Fifty years later, with the ancient culture and ways of the Meriam people buried beneath layers of Christian orthodoxy and western values, a baby was born on the island: Edward Koiki Mabo.

He proved to be a bright and spirited boy, quick to learn languages and with a strong sense of his island heritage. 'They used to belt me every time I talked Meriamish,' Koiki later told his biographer. 'I continued to do so and nearly every day I used to bend over the table for the old fellow to give me a few straps on my backside.' The island's Christian elders regarded him as a dangerous troublemaker and when, aged sixteen, he was caught drinking coconut liquor and sleeping with a woman out of wedlock, they ordered he be banished from his island home for a year.

Koiki found work on the luggers as a deckhand and deep-water diver. The silver-lipped oysters had long been fished out, and the pearlers were now searching the strait for trochus (*Tectus niloticus*), a sea snail shaped like a spinning top, its shell also lined with valuable nacre. The teenager would put on goggles, take a deep breath and dive fifty feet down in search of pearl, just as his grandparents did. It was a grim and dangerous business. The industry was in decline, the market for pearl was collapsing and the money for islander crew was pitiful.

Life on the luggers took its toll and Koiki became ill, ending up in Cairns, an Australian port city in north-east Queensland. 'I'd never seen a village so big as Cairns,' Koiki later wrote. It was a moment of revelation, his own coming of the light, as his eyes were opened to the ways of the wider world. Koiki was outraged to learn that white labourers on the mainland were earning £50 a month, while black islanders received just £17. The mission on Mer, it now appeared, had been educating and training the children of the island for a life of exploitation and servitude.

Koiki walked west along the railway tracks, out into the desert wilderness of continental Queensland, looking for a job and a future. The teenager introduced himself to Australians as Edward, his English name, rather than Koiki, his island name. To survive among the mainlanders, he realised, you needed to speak their language. The guys in the bar called him Eddie and helped him find work on the sugar-cane plantations, a backbreaking hand-to-mouth existence that still bore the scars of the slave trade.

Koiki eventually settled in Townsville, 175 miles south of Cairns, a coastal city dominated by Castle Hill (*Cootharinga*), a pink-granite monolith that offered him spectacular views out to the islands: to Magnetic Island just across the water where Captain James Cook's compass went awry in 1770; to Palm Island in the north, where natives once spoke their own dialect of the now extinct Wulguru language; and, invisible beyond the horizon, marking the boundary between continental and island domains, the Great Barrier Reef.

Koiki met and married Bonita, whose family came from Palm Island. Netta, as everyone called her, was to be Koiki's soulmate for the rest of his life. Like him, Netta saw herself as an islander and had a deep sense of the thread that led back across time to her island ancestors. Together they set up the Black Community School in Townsville, teaching children of exiled islanders the songs, stories and languages of the Torres Strait.

It proved a provocative move, prompting Queensland's education minister to denounce the school's mission as 'apartheid in reverse'.

Koiki was also active in black politics, campaigning for the rights of Aboriginal and islander workers, and reinforcing his reputation as a troublemaker back on Mer. His requests to return to visit his dying father and other members of his family were formally rejected by the Murray Island Council, which feared he would be a destabilising influence.

It must have been agony for this child of the islands to be marooned on the continent, cut off from his family, his village and his land. Koiki took a job as a gardener at the James Cook University, where he would often spend his lunch break in the library trying to reconnect with his ethnic and cultural heritage. He read and reread the *Reports of the Cambridge Anthropological Expedition to Torres Straits*, a volume first published in 1901 and compiled by the British academic Alfred Haddon. In the last days of the nineteenth century, Alfred had spent more than a year among the islanders, determined to chronicle their traditions and ceremonies before they were obliterated from the collective memory by zealous missionaries. To Koiki, the book was like the gasp of air he used to take when surfacing from a deep dive.

Islanders have a sense of living in a confined space, of the sea pressing against the shore. It is a feeling that translates into their relationship with land. The instinct to enclose, to place a boundary around property, appears particularly strong on the British Isles. The eighteenth-century English jurist William Blackstone believed there was nothing that engaged the affections of his fellow man as 'that sole and despotic dominion which one man claims and exercises over the external things of the world'.

The private ownership and control of land has long been central to the social structures of Britain, certainly since William the Conqueror put on his surveyor's hat and ordered the Domesday Book. The British embraced feudalism more enthusiastically than elsewhere in Europe and, along with the widespread enclosure of what was once common land, developed a class and power structure based primarily on who owned what. A person's title was both a legal property document and a status. Walls and hedges were statements of authority. A private garden was the exclusive domain of an elite.

The growth of cities and their suburbs saw the new middle classes demand homes with gardens, modest versions of the great estates of the landed gentry. Owning land and property was, and to some extent remains, an indication of social rank. But it goes deeper than that. It is about sovereignty, the concept which helped swing the Brexit debate and which chimes with the British islander instinct for self-determination, independence and control.

I pull another photograph from the box. This shows the semi-detached house in the Hampshire village of Itchen Abbas where I spent my teenage years. The picture is dominated by the high and thick hedge (*Cupressocyparis leylandii*) that divided our garden from the neighbours', a barrier sufficiently impenetrable to act as wicketkeeper and boundary when playing cricket with my siblings.

The garden had two distinct parts: an ornamental section with herbaceous borders and beds of roses around a well-kept lawn, and a fruit and vegetable patch draped in netting to keep the birds from my dad's prize strawberries. He, like his father before him, was never happier than when pottering about on his plot with a trug and a fork, pulling out invasive weeds and occasionally rattling a fence to ensure the perimeter was secure.

One year, I now recall, the local vicar decided the people of Itchen Abbas should reconnect with their medieval ancestors and 'beat the bounds'. We gathered outside the Plough Inn on Ascension Day and branches of birch and willow were handed out. The priest and churchwardens then led the crowd to the edges of the parish where the youngest were invited enthusiastically to whack the boundary markers placed there. Hymns were sung and psalms recited. 'Cursed is he who transgresseth the bounds or doles of his neighbour,' the vicar intoned.

Beating the bounds was the time-honoured way of ensuring that the extent of the parish was imprinted upon the collective memory. In the days before maps were commonplace, the ecclesiastical courts thought it vital that villagers themselves should know precisely where their territory began and ended.

The annual perambulation of the parish would sometimes involve young boys being thrashed with branches and struck against boundary stones to ensure they wouldn't forget.

The bitter aftertaste from Koiki's 1974 lunch with two senior historians at James Cook University came with the realisation that islanders' shared memories were not enough. Oral tradition was trumped by written documentation. That was how the white fellas had stolen Koiki's island inheritance. The lands on which he and all Torres Strait islanders had always lived, the precious gardens they had always tended, did not, so the map said, belong to them. A century earlier, on 1 August 1879, to be exact, in the name of an empress on the other side of the world and without the knowledge or agreement of those directly affected, the islands had been designated 'terra nullius', belonging to no one, and so, at the stroke of a legislative pen, became the exclusive and absolute property of the Imperial Crown.

In May 1982, however, Eddie Koiki Mabo's name appeared on a document as leader of a group of Meriam lodging a case with the High Court of Australia. They were seeking to establish legal possession of what they had always thought was theirs. It was an extraordinarily audacious challenge, the people of a tiny island taking on the constitutional authority of a continent.

The argument hinged on whether the people of Mer
had a concept of land ownership before the April day in
1770 when Britain, in the form of James Cook aboard
HMS *Endeavour*, planted its flag upon Australia's eastern coast. There was no written paperwork, no formal
system of title or sale, no maps or surveyor reports,
none of the evidence that would normally be essential in
such cases. Ranks of expensive government lawyers with
boxes brimming with documents were lined up to ensure
the matter was dealt with swiftly and decisively in favour
of the status quo.

But the authorities had not bargained for the determination and spirit of the Mer islanders and, in particular,
the indomitable Eddie Koiki Mabo. Thirty-three Meriam
people contributed to 4,000 pages of transcripts, evidence which made the case that there had been clearly
defined territories on the island for centuries. Everyone
on Mer, from each of the many tiny villages, understood
whose garden was whose. The map, though, was in their
heads, not written down.

As barristers studied traditional survey plans, the
Supreme Court Justice Martin Moynihan listened as witness after witness described where the invisible boundaries of ownership lay. 'This is my area from the post
to the almond tree in the centre,' one islander told the
court. 'There is a stone there and from the stone right
down to the beach. On the right side, the post mark the

boundary. From there to that cotton tree right down to the beach, this is my area.'

For three days, the court convened on Mer island itself, hearing evidence from sixteen islanders, many elderly and frail, each with a description of their garden plots and adjacent seas.

Koiki took to the witness box, explaining the position of what islanders called 'Eddie Mabo's portion' by listing the gardens and villages with their tribal names. 'Korkor, Kieu, Utirib, Pairemed, Tar, Niwe, Ugab, Jeum, Nimap.' It was a memorised sound poem of island geography stretching back seventeen generations. 'Baier, Maskep, Aeum, Sager, Las, Dam, Miear, Dure, Kebisaurem, Manini, Sebeg,' Koiki continued his ancient chant. 'Saurem which goes past the road into the rain forest to the bamboo patch way up into Kirr where there are heaps and heaps of rocks.' The detail was compelling. 'Lewag, Umargiz, Kosmed, Mabadmop, Parbid, Semar, Ebir and Ulag together with the sea, seabed and stone fishtrap extending from the land at Las to the fringing reef and the Great Barrier Reef.'

The photograph box is full of images of the places where I grew up. The outline of buildings, the curve of paths, the layout of the gardens, the shapes awaken memories of days gone by. We all have a homeland, a place with a meaning, where the physical meets the personal. Homeland is about legal status and right of abode, but it is also about connection through stories and experiences. We know where we belong. The perception might be more acutely felt by those on physical islands for whom the sea marks a clear territorial junction, but everyone has at least a sense of where home turf begins and ends.

The eighteenth and nineteenth centuries saw the expansionist colonial powers of Europe, notably the British and the French, use the power of legal documentation to inscribe their jurisdiction over the world. Maps and charts, treaties and title deeds all became the stuff of sovereignty, drawing the lines on the globe that decided who owned what. Where the rightful possession of a territory was in doubt, cartographers might simply leave it blank, rather in the way Ptolemy had marked unknown land as '*Terra Incognita*' on the maps in his work *Geography*. A blank really meant ripe for the taking.

'Now when I was a little chap I had a passion for maps,' the character Charles Marlow recalls in Joseph Conrad's novel *Heart of Darkness*. 'At that time there were many blank spaces on the earth, and when I saw one that looked particularly inviting on a map (but they all look that) I would put my finger on it and say, "When I grow up, I will go there."'

Detailed survey maps were among the archive of official documents which gave legal weight to land ownership in English law, but this reliance on technical plans obscures an older but equally valuable method of surveying property in Britain: metes and bounds.

Before cartography took over, the accepted way to define an area of land was to explain its place in the local geography. Metes were straight lines running in a particular direction between two fixed points, perhaps a tree and a barn, while bounds were more general descriptions of how a boundary related to the terrain, running along a wall or a river or a roadway. For centuries, British surveyors would identify the outline of a parcel of land using a running prose style, working around the edges in sequence before returning to the start point.

What this method lacks in precision it makes up for in poetry. I hear echoes of the Lapita people in their canoes, chanting the directions to distant shores, as well as the garden mind maps of the Torres Strait islanders, of course. Reciting a mantra of metes and bounds is surely truer to the psychological relationship people have with their land, more in tune with landscape and a personal sense of homeland than any technical drawing.

On 3 June 1992, ten years after the case had begun, Justice Martin Moynihan prepared to announce the decision of Australia's High Court in the case of *Mabo v. Queensland*. As the arguments had dragged on, Australia's big farming and mining companies had become increasingly nervous, aware of how businesses worth billions operated on land that had also once been declared *terra nullius* and claimed for the Crown.

'Gardening', Martin began, 'was of the most profound importance to the inhabitants of Murray Island at and prior to European contact.' The judge explained how the gardens of Mer actually transcended common law notions of property or possession. They were their homelands, an integral part of the social, economic and ritual life of the island. 'Prestige depended on gardening prowess,' he continued, 'manifest in the show gardens and cultivation of yams to a huge size.'

One can only imagine how the corporate lawyers must have felt as they listened to a judge from Australia's most senior court explain the importance of competitive vegetable growing for the people who lived at the end of the archipelago.

'Considerable ritual was associated with gardening, and gardening techniques were passed on and preserved by these rituals,' the judge went on. 'The boundaries of their traditional lands were likely to be long-standing and defined.' The conclusion, he announced, was inevitable. 'The Meriam people are entitled as against the whole world to possession, occupation, use and enjoyment of the lands of the Murray Islands.'

The share values of some of Australia's richest corporations plummeted on the news, but for the islanders who had travelled almost 2,000 miles to be in the Canberra courtroom, there was elation. Their only sadness was that Eddie Koiki Mabo was not there to witness his victory over the white fellas who tried to steal his garden.

Five months earlier, Koiki had died from cancer, his wife Netta at his bedside. His last words were: 'Land claim.'

I travelled to Murray Island for the BBC, just after the court ruling, to report on the legal and political implications of what had become known as the Mabo judgment. What touched me, though, was the spiritual dimension of the story, the sacred bond I saw between islanders and island. To the lawyers in the High Court, the argument was about territorial ownership. To the Mer people, it was about a deep personal relationship with the landscape. There was nothing in the plots they proudly showed me that mirrored the privet hedges of the neat gardens in Balmoral Drive or the towering curtains of leylandii in Itchen Abbas, no fences to rattle or boundary stones to beat. But I recognised the way the shapes and patterns of the land had become tattooed upon their consciousness.

One elder had taken me out in his boat, beyond the reef. He pointed at the island and said, 'What do you see?'

'I see Murray,' I replied.

'What is the shape?' he pressed me.

'I don't understand,' I said.

'It is the narwhal,' he explained, pointing to the sinuous outline of the island with its tusk-like finger of coral sand stretching from steep forested hills. 'Do you see? The Christians came, but we will always worship the narwhal.'

Every grain of sand on the beach, every leaf on every branch on every tree, every feather on every bird had a blessed place in the hearts of the native people. They were not owners of land but custodians of paradise. Each bore a heavy responsibility to protect and nourish the earth beneath their feet and the waters that lapped against their shore, to pass the island to the next generation as they had received it from the last.

The physical and the spiritual, island and islander, had become merged like sand and sea on the beaches of Mer.

CHAPTER 14

UTOPIA

In Pursuit of Paradise

The English county of Essex trickles into the yawning mouth of the River Thames in a filigree of shining creeks and gullies and rivulets. Oystercatchers (*Haematopus ostralegus*) and redshanks (*Tringa tetanus*) dip their long bills into the brackish waters where land meets sea, as flocks of dunlins (*Calidris alpina*) swoop against a big sky and over meadows of eelgrass (*Zostera marina*) and samphire (*Salicornia europaea*). There is an energising openness to this landscape, a relief from the claustrophobic intensity of the mega-city upstream.

I have decided to take Pangaea on a day trip to the estuary, more precisely to Canvey Island, a small patch of reclaimed salt marsh, hunkered below sea level on the north bank of the river. She sleeps in my pocket as the train chugs its way out of the capital, along the tracks built to take working-class families away from London's smoky East End to play on the sands at Southend-on-Sea. We get off at Benfleet and take a bus onto Canvey itself.

Visitors might struggle to realise they are on an island at all. It requires a map to explain how the meandering channels of Holehaven Creek and East Haven Creek contrive to separate this toe of alluvium from the body of Essex. But Canveyites, mostly families of migrants from east London, are immensely proud that visitors must cross a bridge to reach their island haven.

I first came to Canvey for the BBC after the 2011 census revealed that the area was the most English place in England. Responding to a question asking people about their national identity, eight out of ten people on the island had ticked the box marked simply 'English', a higher proportion than any-where else in the country.

I had spent the evening in a sports bar, the regulars draped in the cross of St George as they watched their national team on the giant screen. One man in his seventies, wearing a knot-ted red and white handkerchief on his head, told me why he regarded himself as English rather than British. 'I don't think Britain is Britain any more,' he said sadly. 'I have always been English. I was born English.'

I stroll along the esplanade, following the three-kilometre concrete barricade that attempts to keep the sea out of the island. Cottages crouch behind the wall, their view of the open estuary completely obscured by the high flood defences. Islanders are faced with a daily reminder of external threat, the risk of being swamped by an incoming tide. If people struggle to see out, they tend to look in.

Many people moved to Canvey, I suspect, because they

shared similar values to those already on the island; proud to be working class, to be white and to be English. But islandness plays a part in concentrating those attitudes, creating an echo chamber that reaffirms their worldview. Canvey has become a sanctuary, a retreat from the challenges and complexities of urban life. The residents regard it almost as their private Eden, their demi-paradise.

As a boy, Matti Kurikka would stand on the shore of the Gulf of Finland and glower across the bay at Kotlin Island (*Котлин*), the base for Russia's Baltic Fleet. He could see the heavy Krupp guns on the ramparts of sheer walls, a display of imperial force aimed at ensuring the land of the Finns would remain absorbed within the detested Russian Empire.

Three decades later, just before Christmas in 1901, Matti stood on another shore, on Broughton Strait in south-western Canada, staring across the cold waters to Malcolm Island, an uninhabited lump of not very much. To Matti, though, the place was perfect, physically and socially isolated from the adversities of the mainland. Oar in hand, he believed he had found his Garden of Eden.

Matti had always been a dreamer. Son of a wealthy Finnish farmer, he rebelled against his comfortable upbringing and dropped out of university to devote himself to a life of radical socialism. He soon became an acknowledged leader in Finland's labour movement as editor of *Työmies* (*Worker*), a Helsinki newspaper with strong views on communal ownership, workers' rights and women's equality.

His career had barely got started, however, when Tsar Nicholas II passed a decree which sought to eradicate the last vestiges of Finnish values and sovereignty from the territory. The policy of Russification (*sortokaudet*) saw the press subjected to strict censorship, replaced Finnish with Russian as the language of administration and abolished Finland's postage stamps. Eight definitives, from the 2 penni in grey to the 1 markka in violet and bearing the Finnish coat of arms with its crowned lion, were replaced with stamps from 2 kopeks to 7 roubles, bearing Cyrillic script and the double-headed eagle of the Russian Empire.

In 1899, Matti fled his homeland, ending up in Canada after some Finnish migrants labouring in the dreaded coal mines of Vancouver Island had contacted him asking for help. His answer to their plight was the creation of a new and perfect society, a community free from the exploitative demands of robber barons and Russian imperialists. Malcolm Island had been purchased as the location for a socialist utopia.

On a mid-winter day, Matti and four followers rowed across the bay to a spot on the island they had agreed, in a properly constituted vote, to call *Sointula* (Place of

Harmony). Here was virgin territory, a blank canvas, a location where they could start from scratch, to build a commune founded on the values of consensus, equality and free love.

Even if the frontier is but a network of shallow creeks, the insularity created by an encircling boundary can still generate a profound emotional response. Canvey Island's physical estrangement from the mainland, however subtle, has encouraged a conviction that this protected corner of Essex has a unique and special character, a charm not properly recognised by those on the other side of the water line. Over time, the joys of separation have darkened into feelings of marginalisation and resentment.

Dave Blackwell, born and bred on the island, was particularly dismayed by the way his beloved Canvey was being treated, and so, in 2004, he founded the Canvey Island Independent Party (CIIP). 'During the 1950s and 1960s, Canvey Island was a lovely place to live; clean and safe. Where has all of that gone?' Dave asks on the official party website. 'Well, it seemed to have started when we joined the mainland.'

A reorganisation of England's local government in 1974 saw officials ignore the islandness of Canvey, its district council subsumed into a new borough, Castle Point, that also included the mainland communities of Benfleet, Hadleigh and Thundersley. There was quiet fury that Canvey's perfect island status had been desecrated by an arrogant Whitehall elite which did not seem to understand or care.

The idea of the perfect island, a place immune to the faults and disappointments of the wider world, has captivated imaginations for thousands of years. In the late fourth century BC, the Greek philosopher Euhemerus wrote about the island of Panchaia in the Indian Ocean, a 'very splendid land' possessing three prosperous cities, a forest of spice trees and a wealth of gold and silver. He described it as a rational island paradise, combining the best of human logic and order with the perfection of the divine.

That mix of the earthly and the heavenly was also to be found in Plato's story of Atlantis, the island utopia whose founders were half god and half mortal. According to accounts, Plato explained how this ideal society was constructed as a series of concentric islands, each filled with exotic creatures and precious metals, separated by wide moats and linked by a canal. The inhabitants were moral and spiritual, leading perfect lives. However, Atlantis contained the seeds of its own destruction, the hubris of corruptible people. In time, they became greedy and decadent, and eventually the gods decided they must punish the fallen people of paradise. In 'one terrible night of fire and earthquakes', Atlantis sank into the sea for ever.

The word 'utopia' was coined by Sir Thomas More in 1516, first appearing in the long Latin title of a work which translates as *A little, true book, not less beneficial than enjoyable, about how things should be in the new island Utopia.*

Sir Thomas explained to his readers how Utopia was originally a peninsula on the continent of Utopus, its isolation manufactured by the digging of a channel that 'brought the rude and uncivilised inhabitants into such a good government'. Islandness was a pre-requisite for the creation of a perfect society. But, of course, Thomas's 'true book' was a work of fiction. His term 'utopia' was derived from the Greek words for not (*ou*) and place (*topos*). There, in plain sight, was Thomas's point. Utopia did not exist. It was nowhere.

The idyllic islands of Utopia and Atlantis felt tangible to Matti Kurikka, however, as he set about building his private paradise on Malcolm Island that winter's day in 1901. As a follower of Theosophism, an early new age movement founded by Russian mystic Helena Blavatsky, he had been taught that Atlantis was the true location of the Garden of Eden, and that its population of cultural heroes perished when they abused their psychic and supernatural powers. Theosophists, like others before them, were searching for understanding and enlightenment in the flotsam and jetsam of a shattered island utopia, a civilisation which perished because of the flaws of mortals.

Matti should, perhaps, have paid closer attention to the moral of the story. Within every utopia there was a dystopia. Paradise was an unattainable dream because the perfection of the virgin beach was lost as soon as the first footprint was imprinted upon the sand, as soon as the boundary was crossed.

The administrative changes that sought to deny Canvey its islandness coincided with the UK's decision to join the European Economic Community, later the European Union (EU). For decades, both events would gnaw away at Canveyites' sense of lost island sovereignty. 'Residents say they are fed up with being ruled by mainland Tories,' Dave Blackwell wrote to voters as he campaigned for a more independent Canvey. 'The mainland Tories, which rule the Council, see Canvey as the place to develop to save their own mainland green belt.'

The Canvey Island Independent Party was initially regarded with amusement by the power blocs dominating local politics in south Essex. But within three years, Canvey Island had replaced all but two of its seventeen Conservative and Labour borough councillors with representatives of the CIIP, and the party has dominated island affairs ever since. That desire to regain sovereignty, to take back control, was evident in islanders' attitude to Britain's membership of the European Union too.

On a cold February day in 2015, reporters, television crews and politicians gathered in the foyer of the Movie Starr cinema on Canvey's Eastern Esplanade. Purple and yellow boxes of popcorn were handed out to confirm that Canvey Island had been chosen by the United Kingdom Independence Party (UKIP) as the spot to launch their election manifesto. Headed by Nigel Farage, UKIP was confident its demand for Britain to exit the EU and its brand of popular nationalism would chime with the values of Canveyites.

When the Brexit referendum came the following year, 73 per cent of people in Canvey's constituency voted for the UK to leave the European Union, one of the most emphatic results anywhere in the country.

I walk over the road from the Movie Starr cinema, across a car park and up an embankment to the concrete seawall belatedly installed after terrible floods killed fifty-nine people on Canvey in 1953. Sheltering beneath the barricade on the estuary side is a line of wooden benches, many with plaques remembering islanders who had once liked to sit there and watch the boats go by. I pull Pangaea from my pocket and wonder at her power.

Canvey calls to those who seek refuge from the global metropolis upstream. In June 2016, just as the UK voted for Brexit, six houses on the island were bought by members of London's Hasidic community, a Jewish sect that conforms to a strict form of orthodox Judaism incubated in the ghettos of Eastern Europe. Insularity shaped and intensified the group's distinctive identity. Struggling to find affordable accommodation for their large families in north London and wanting to shelter their traditional way of life from the modern liberalism of the capital, they saw in Canvey a place to relocate their community. Within three years, seventy-five Jewish families had made the move east from Stamford Hill to the island on the estuary. The press called it an exodus to the Promised Land.

There had been concerns of a culture clash, that native Canveyites would resent the strange incomers to their island. But that is not the story.

On the night of 22 December 2019, most of Canvey was decorated for Christmas while the Jewish families were marking the start of the festival of Hanukkah. A menorah had been left burning in the religious school (*yeshiva*), a former college that had become the Hasidic community's social centre. It is thought the candle may have fallen. The fire swept through the building destroying almost everything, but it also illuminated the respectful relationship that had developed between the two traditions which now shared the island. Locals and newcomers worked together to salvage what they could from the ashes. 'The support from the islanders was overwhelming,' one Hasidic representative told reporters.

The two communities are very different, but they recognise a common desire to protect their culture and way of life. Both groups can trace their stories back to the struggle to survive in London's East End, and both reached the same conclusion about Canvey and its precious islandness. When disaster struck, they were there for each other.

By 1903, Sointula had a population of 238 Finnish inhabitants, including eighty-eight children. But the commune's story was one of extreme hardship, disharmony and tragedy. In January of that year, a terrible fire consumed the community centre, killing eleven people and destroying many of Sointula's records and supplies. It was a bitter blow that must have appeared to have symbolic significance to the fledgling settlement, struggling to feed itself from the thin profits of logging and fishing industries. Matti's brand of radical socialism had also proved too much for some of the residents. 'Marriage and love are two different things, just as the church and truth are two different things,' he had suggested. But his alternative proposition, a philosophy of free love and communal childcare, was not embraced by the community in Sointula, particularly the women. The arguments became increasingly fierce and, three years after rowing ashore, Matti sailed away from Malcolm Island for the last time. Those left behind blamed their plight on unrealistic idealism when what was needed in the circumstances was practical common sense.

Matti was not finished. He tried to set up another utopian Finnish colony east of Vancouver, but this time without the inconvenience of women. The bachelor commune lived together in one large cabin, an island of simple celibacy where a man's dreams could be dreamed.

But his followers soon grew disillusioned with a lifestyle of denial and, while Matti was away on a lecture tour, decided to put their celibate days behind them. They wrote to their erstwhile leader advising him that he need not bother to come back.

As I saunter along Canvey's promenade, Pangaea nestling in my coat pocket, I see an illuminated sign promising 'Fantasy Island Amusements'. A stall by the entrance sells candy floss, bright pink clouds of spun syrup that disappear almost to nothing on contact.

Humankind has always been searching for the path to paradise. We fool ourselves into believing that utopia will be an island of our own righteousness, created in our image and governed by our values. 'My gaff, my rules,' as pub landlords are inclined to say. If we could only take complete control and exclude the troublemakers, then we would build our promised land. Except we know deep down that this is to believe in a fantasy island, a confection which cannot survive human contact. Like the people of Atlantis, we are all flawed, and any vision of utopia is too. Should the terrible night of fire arrive, we may need each other.

CHAPTER 15

NATIONHOOD

The Divisions Within

No one was sure who or what had started it, but people had their suspicions. Armed mobs from the Istanbul Muslim gang and the rival Texas Creole gang were roaming the streets, both sides seeking revenge for a killing they blamed on the other. On one level, it was simply an escalation of the familiar struggle for control of the slums of Port Louis, a turf war over the capital's sex, drugs and gambling rackets. On another, it was a warning shot in a bigger battle for control of Mauritius, just as the island was counting down to its long-awaited Independence Day.

The *'bagarre raciale'* ('racial scuffle'), as the riots of January 1968 were to be remembered, turned neighbourhoods where people from African and Indian heritage had coexisted for a hundred years into segregated communities aggressively competing for political advantage, as the island took its first tentative steps as a sovereign nation.

Some suggested the man known as *Le Roi Creole* (King of the Creoles) lit the fuse for the rioting. Gaëtan Duval was the flamboyant leader of the populist *Parti Mauricien Social Démocrate*, a right-wing group that had helped stir up anti-Indian feeling on the island with provocative racist slogans: '*Malbar nous pa oulé*' ('We don't want Indians') and '*Envelopé nous pa oulé*' ('We don't want to wear Indian clothes'). It was crude and offensive, but it played to the fears of both rich whites and poor blacks that independence would hand control of their island to the largest of the ethnic groups, the Hindus.

It is thought twenty-five people had lost their lives in the rioting before the island's governor called for help from the British Army. Troops from The King's Shropshire Light Infantry arrived from Malaysia to restore order, hardly an auspicious sign as Mauritius prepared to show the world that it was ready to cut its ties with the old colonial administration.

One might think that islands are designed for nationhood. If, in the Irish historian Benedict Anderson's celebrated phrase, nations are 'imagined communities', islands surely have an advantage. They can simply point to the coastline as physical evidence of where territorial sovereignty begins and ends.

Boundaries are clearly defined along that enchanted stripe where land and water kiss, drawing unique and recognisable shapes against the blue-black sea. Continental nations often struggle to describe themselves geographically, their manufactured frontiers shifting back and forth over bloody battlefields, blurring into landscapes that seem reluctant to recognise them. Islands have a natural integrity that mainland states can only envy.

But, of course, nationhood is not about geology or maps. It is a state of mind, a shared sense of belonging. An island perimeter may amplify that spirit of community, or it may do the opposite, intensifying the differences between the cultures and traditions, the motivations and ambitions that reside within. History, sadly, is littered with the latter.

I reach up to my bookcase and pull down a dark green volume with 'The Strand Stamp Album' embossed in gold on the front. It is a relic from my childhood representing untold hours spent with tweezers and magnifying glass, hinges and catalogues, identifying and organising the stamps of the world in their correct place within an alphabetical list of countries and other issuing authorities. I can feel Pangaea's watchful gaze as I turn the pages looking for something.

There it is. At around the age of eight, I must have licked the hinge on the back of that small green stamp and stuck it down carefully among the set, oblivious to how the design would imprint itself on my memory so that when I think about the birth of island nations, I reconnect with this scrap of paper gummed to the page headed 'Ireland'.

In 1922, James Ingram from Glasnevin in Dublin won £25 for the artwork, unaware that his elegant drawing would become a symbol of the fledgling Irish Free State. The postmaster general of the new nation of *Éire* (Ireland) had arranged a public competition to find a suitable design to replace a hangover of British rule, the orange 2d postage stamp used for standard letters, with its profile of a bearded King George V enwreathed by oak and laurel leaves beneath the Imperial Crown. For a time after independence, the Irish Post Office had resorted to printing the words '*Rialtas Sealadach na hÉireann 1922*' ('Provisional Government of Ireland 1922') across the monarch's head, but nationhood required its own symbols of identity and postal officials were keen to start producing stamps that expressed the optimism of the moment.

James's submission replaced the cartouche of leaves with a Celtic arch decorated with knots, scrolls and shamrocks, substituting His Majesty's head for an image of equivalent potency: the outline of the island of Ireland.

The judges at the Post Office must have known it would be seen as a statement, not just of national pride but territorial claim. The map included the six counties of Northern Ireland that were not part of Éire's self-governing dominion. Exactly a year after the peace treaty had been signed, an agreement that created a twisting land border between the new Irish nation in the south and retained British territory in the north, the 2 *pingin* (penny) green was issued, emphatically illustrating the complete and total islandness of Ireland. It was an image that would be fixed to millions of letters and cards posted across the British Isles over five decades. James's island design was only withdrawn from sale in 1968, making it one of the longest-running stamps in the world.

When people refer to the Emerald Isle, it is that shape of the island of Ireland that comes into my head, a pleasing outline that looks somehow balanced and logical, as though drafted by an expert hand. But like the political border winding its provocative way across the landscape, almost invisible to those without local knowledge of where north meets south, beneath the green sward are hidden rocks which also tell a story of opposing forces and explosive violence.

Around 600 million years ago, the place we now call Ireland was in two parts, separated by an ocean. One half was on the continent of Laurentia, preserved today in parts of North America, and the other on Gondwana, the supercontinent which provided the tectonic building blocks for most of the land on the planet. Over the next 150 million years, these huge hunks of terrestrial crust inched slowly towards each other until, with chaotic inevitability, they collided, an impact that made the world shudder and buckle.

I glance at my muse dozing on her couch. She seems to know what is coming next.

The various pieces traversing the oceans of the globe appeared drawn to each other, with the small patch of conjoined terrain which would later become Ireland swept northwards from the tropics, often submerged by warm coral-rich seas, until incorporated into the very heart of the vast island of Pangaea, its mixed parentage forgotten in the immensity of the congregation.

As Pangaea gave birth to the continents and islands as we now see them, Ireland travelled upon a particularly violent road to independence. Around 60 million years ago, magma from deep below the earth's surface forced its way up through the compromised scars of its geological past, spilling molten rock onto the ground in a lake of lava that cooled rapidly to form a unique landscape of hexagonal basalt columns. The Giant's Causeway is a wonderful legacy of the strange upheavals that created Ireland.

Looking down from the International Space Station on a cloud-free day, photographs show the carpet of green that covers the familiar outline of Ireland, obscuring the geological divisions and rifts beneath. The island integrity we perceive is really a patchwork stitched by time: remnants of desert dunes and coral reefs, forests and lakes, ocean currents and seismic paroxysms.

Lying on her couch, Pangaea seems to be reminding me how islands can appear deceptively cohesive, their geography suggesting unanimity and rationality. But scratch the surface and one will encounter the wounds of yesterday and the frailties of today.

Huge volcanic eruptions some 8 million years ago saw Mauritius bubble and steam from the depths of the Indian Ocean. Along the arc of the undersea Mascarene Plateau, dozens of islands broke the surface of the sea, some to sink back down, others still managing to keep their heads above the water. From Réunion in the southwest, across to Rodrigues in the east and north to St Brandon (*Cargados Carajos*) with its transient collection of sandbanks, shoals and islets, Mauritius was among a group of small islands geologically detached from distant Madagascar and the African continent beyond.

Extreme isolation allowed the development of a unique ecosystem. Without terrestrial mammals, birds and bats and reptiles evolved in wonderful ways. Giant tortoises and enormous lizards roamed forests filled with soaring ebony trees and strange plants. In the air were parrots and parakeets, fruit bats and owls found nowhere else on the planet. The most celebrated resident of this exclusive tropical club was the dodo, a large and flightless member of the pigeon family which pottered contentedly around the wooded coasts of Mauritius.

It was a fragile island paradise, however, that could not survive contact with voracious human beings and the vermin which travelled with them. Descendants of the Lapita people in their outrigger canoes and Arab sailors in their dhows may have stopped on their way to somewhere else, but it was the arrival of the Europeans that really did for the dodo and some of its extraordinary compatriots. The Dutch established the first settlement on Mauritius in 1638, bringing rats, macaques, pigs and cats. The trusting dodo was on the edge of extinction within twenty years. By the time the French took control of what they renamed Isle de France in 1715, the island's giant domed and saddle-backed tortoises were all dead.

White colonisers reimagined paradise as an unrestrained engine of profit. Ancient ebonies were cleared for sugar-cane plantations to be worked by thousands of slaves shipped in from Madagascar and from across the continent of Africa. The French surrendered the island to the British in 1810 but on terms that allowed for business as usual. The plantation owners kept their land and their language, as well as ownership of around 65,000 enslaved Africans. Even the abolition of slavery in 1834 could not stop the money-making machine that Mauritius had become. The entrepreneurial planters simply looked east rather than west and actually expanded their operations by employing vast numbers of indentured workers from India, forced labour not technically outlawed as slavery. Some 200,000 Indians arrived on the island over the next few decades, becoming easily the largest ethnic group in Mauritius.

A small island that was uninhabited until the end of the sixteenth century had, by the end of the nineteenth, a population of more than 370,000, a diverse mix of Indians, Africans, Europeans and even a Chinese community. In terms of religion, culture, status and identity, there was little to unite the peoples of Mauritius apart from the fact that they had all arrived by boat to service the sugar trade.

The supremacy of the white oligarchs, who controlled the wealth that flowed like black treacle from the multicoloured Mauritian soil, began to be challenged in the early twentieth century by a new middle class of Indian Hindus. Mahatma Gandhi had encouraged the island's rural Indians to get educated and get active while on a trip to Mauritius in 1901, and the next sixty years saw their influence flourish as the prospect of independence became more real. But the shifting plates of the island's power tectonics threatened the stability of a new nation emerging from deep colonial waters. It was small wonder that the British governor, Sir John Rennie, called the army for help when the body count of Port Louis's slums entered the dozens in January 1968, just seven weeks before the ceremony to mark the birth of an independent Mauritius.

As it turned out, the rain that greeted the appointed day of 12 March had given way to bright sunshine by noon, when the Union Jack was lowered, and the rainbow flag of Mauritius hoisted in its place. The ceremony at the Champ de Mars in Port Louis heard the national anthem performed for the first time. 'Glory to thee, Motherland,' the choir sang. 'Around thee we gather as one people, as one nation.' It was an anthem of unity, but Gaëtan Duval was not there to hear it. The King of the Creoles stayed at home eating 'cari poule et farata' ('chicken curry and roti bread'), as a firework display lit up the harbour, echoing the explosive birth of the island some 8 million years earlier.

British soldiers patrolled the streets of the capital that

night as a nervous people wondered how independence would play out. Could this small island, fragmented by cultural rivalries, reimagine itself once again, free from its colonial masters? Thousands didn't wait to find out the answer, boarding planes and boats to start new lives in Australia, Canada, South Africa, France and Britain.

Pangaea manages to doze at the bottom of my overnight case as it trundles noisily out of the airport terminal. I have decided to take her on a work assignment, to the Emerald Isle.

The peace walls of West Belfast snake through Northern Ireland's capital, adorned with bright street art and murals. I disembark from a bus that pulls up to let cheerful tourists write their messages of hope and love on the bricks. But the walls are not symbols of peace. They are structures of fear. Their purpose is not to celebrate peace. It is to keep the peace.

It is more than two decades since I was in this city to report on the signing of the agreement that was supposed to mark an end to murderous sectarianism on the island of Ireland. I remember the optimism and excitement about how communities tortured by the Troubles might spend their peace dividend. Returning, I am struck by how divided and nervous Belfast remains. The walls erected to reduce inter-communal violence are not only still here, they are longer and they are higher than ever. By 6 p.m. each day, metal gates between the

Protestant Shankill Road and the Catholic Falls Road have been clanged shut.

I check into a hotel and unpack, placing Pangaea on the table. Many people have attempted to unravel the causes behind Ireland's apparently intractable divides, but the ABC of islandness must surely play its part. The theory of 'amplification by compression' suggests isolation cranks up the contrast, emphasising inequalities and magnifying grievances.

It is no coincidence that the pressure cooker of pandemic lockdown exploded into the Black Lives Matter protests of 2020, frustration boiling over into anger on city streets. In Britain, symbols of colonial oppression were toppled and thrown into the river. Incarceration reminds us of what freedom looks like.

Nelson Mandela spent years mapping out the 'long walk to freedom' in his tiny cell on Robben Island. 'The hunger for my own freedom became the greater hunger for the freedom of my people,' he said. 'The chains on all of my people were the chains on me.'

The most important written text of the American civil rights movement was composed by Martin Luther King while locked in a 'narrow jail cell', the prison walls reminding him of his birthright of freedom. 'Injustice anywhere is a threat to justice everywhere,' the great preacher argues in his 'Letter from Birmingham Jail'. 'We are caught in an inescapable network of mutuality, tied in a single garment of destiny.'

Isolation encourages introspection. Confinement focuses

the mind. The crash and spray of an island shoreline symbolise the struggle for freedom.

The Trinidadian novelist V. S. Naipaul visited Mauritius in 1972 and described 'an agricultural colony, created by empire in an empty island and always meant to be part of something larger, now given a thing called independence and set adrift, an abandoned imperial barracoon, incapable of economic or cultural autonomy'. The population, he wrote, was 'obsessed with the idea of escape' from a land of 'sugar cane and sugar cane, ending in the sea, and the diseased coconut trees blighted by the rhinoceros beetle'.

The question that hung in the air was stark: without imperial imperative, what was the point of Mauritius? Indeed, what sort of future could there be for many little islands dotted around the globe as they took a knife to the colonial umbilical? They were societies and economies hardwired to do one thing which now had the delicate task of reconfiguring themselves for a different purpose.

The world looked on, expecting that Mauritius would prove to be doomed like the dodo, trapped at an evolutionary dead end, unable to adapt. With sugar accounting for almost all exports and one in five of the adult population workless, prospects looked decidedly unpromising.

But there was a character to the islanders developed by generations of privation, a resilience and an ingenuity that many had not bargained for. The economy began to stretch its wings, moving into textiles and tourism, financial services and manufacturing, new sectors taking off on an island which wanted to show the world that, against all the odds, the sovereign Republic of Mauritius could really fly.

Diversifying from the monoculture of cane was one challenge, but unifying the many cultures and creeds living under the multi-coloured flag was quite another. The divide-and-rule policies of the slave owners and the social hierarchies of European and Indian class systems had emphasised distinct ethnic and religious identities on the island, communities separated, in the words of Mauritian author Malcolm de Chazal, by 'freemasonries of blood'. The unifying potential of the island's geography was negated because Mauritians, almost without exception, felt they were connected to a cultural backstory beyond the shoreline, from a continental motherland in Europe, Africa, India or China.

Without an indigenous culture of its own, the island had never developed a recognised tradition of Mauritianism – a repository of stories, craft, music or dance that all islanders could claim and share. There was a widely spoken local language, Mauritian Creole (*kreol morisien*), a hybrid developed over generations as a practical form of communication on a colony where no native tongue predominated. But English remained the language of law and French the language of the media.

As Mauritius took its first faltering steps as a sovereign state, there appeared to be little upon which to build a shared national identity. Instead, it was decided the aim should be '*le vivre ensemble*' ('living together'), where communities would rub along peacefully, each retaining a distinct and cherished character. The policy was legitimised with a voting system that tried to ensure representation of minority factions by rewarding the 'best loser' in elections. Communalism became embedded in the politics and democracy of the island.

For all the anthem's hopes for one people and one nation, Mauritius had chosen to pursue what politicians described as a 'fruit salad' or 'cultural mosaic' approach, the task of constructing a shared Mauritian identity left for another day.

I notice a brochure on the desk in my hotel room, advertising the tourist attractions of Northern Ireland. On its cover is a striking photograph of the Irish coast, a small circular building on top of a sheer cliff staring out defiantly at the Atlantic below. Pangaea prompts me from her couch, apparently expecting me to respond to this island scene, and so, the next morning, I catch a train to Castlerock, rattling an hour and a half north-west to the beaches of County London-derry (County Derry).

Perched on its basalt throne and framed by a glorious sky, it is obvious why Mussenden Temple is one of the most photographed structures in all of Ireland. Positioned as if to keep watch over the perimeter, the building now finds itself in danger of toppling into the sea as the ocean gnaws away at the land beneath it. But the calm logic of its architecture gives it an air of serenity above the fretful boundary where an island's identity begins and ends.

Commissioned by the extraordinary Earl Bishop of Derry, Frederick Hervey, in the 1780s, the building is an exercise in symbolism, full of hidden meanings and clues. (Frederick, incidentally, was the younger brother of Augustus, the admiral after whom Captain James Cook named an archipelago in the South Pacific, mentioned in an earlier chapter.)

A Latin inscription around the outside of the temple reads: '*Suave, mari magno turbantibus aequora ventis e terra magnum alterius spectare laborem*' (''Tis pleasant, safely to behold from the shore, the troubled sailor, and hear the tempests roar'). It is a line from the Roman philosopher Lucretius's *De Rerum Natura (On the Nature of Things)*, an odd choice of inspiration for an Anglican bishop one might think. *De Rerum Natura* was scorned as an atheist manifesto by early Christians, with its strident arguments against absolutist religious dogma.

Constructed as a library in the classical style of Rome's Temple of Vesta, Frederick kept a fire burning day and night in the basement to prevent the briny sea air from damaging the books stored there. The temple represents enlightened thinking, learning and understanding, resolute in the face of irrational chaos.

In the floor is a partially hidden trapdoor leading down to a crypt where Catholic masses were held in secret. Frederick appears to have been using the architecture to make a point about religious tolerance, and it is probable that the building's dedication to Vesta was a nod to her syncretic qualities: she is the counterpart of the Greek goddess Hestia and an example of how the Romans absorbed deities from other traditions into a single pantheon. She is a cross-cultural symbol of hearth and home.

Looking at the building, obdurate and defiant on the shoreline, I can sense Frederick's passion for the islandness of Ireland. He has constructed a statement on integration and parity of esteem which mirrors his own political activities at the time. As the temple was taking shape in his mind, Frederick threw himself into the Irish Volunteer movement, demanding Catholic emancipation and independence for Ireland. He attended the Dungannon Convention of 1782, a church meeting which pressured the British government into giving greater autonomy to the Irish Parliament. A year later he was in Dublin, dressed in full ecclesiastical purple in an open landau being pulled by six carriage horses in purple and gold trappings with young parsons acting as outriders. Frederick had hoped to be the grand figurehead for a new tolerant and united Ireland, a nation true to its island integrity and with sovereign status. But tribal politics drained the energy from that dream. Frederick lost interest and departed for continental Europe. On the last day of the year 1800, with the earl bishop estranged from his family and his Irish estate in a house near Rome, the Kingdom of Ireland was abolished.

I pull Pangaea from my bag and place her in the dawn-facing doorway of Vesta's temple. We cannot stay long: a wedding is due to be held in the building and the families are gathering. Almost two and a half centuries after being built, the temple is actively doing what its architect intended, bringing islanders from different backgrounds together in love.

Joseph Réginald Topize was born in the summer of 1960 among the cheap four-wall cement houses and backyard pigsties of Roche Bois, a dirt-poor district of St Louis regarded as the crucible of Creole identity in Mauritius. Much of the neighbourhood had been built in a hurry, emergency state accommodation for hundreds of desperate families left homeless by Cyclone Carol, a storm that had ripped Mauritius apart in late February of that year.

By the time of the island's independence, eight-year-old Joseph was to be seen doing odd jobs in a neighbourhood that had become a ghetto for the poorest descendants of African slaves. Like the island, the young boy was showing remarkable economic resourcefulness, but as he roamed the slums, Joseph was also searching for an identity that made sense of both his black heritage and his Mauritianism.

Over the next three decades, Joseph found a way to express himself through music. Inspired by the traditional 'sega' slave songs of his island and the rhythms of Jamaican reggae, he created 'seggae', a musical fusion that represented the polyphony of his Mauritian and African heritage. Seggae was hugely popular among the Creole youth of the Mascarene Islands, connecting them to the struggles of the wider black slave-island diaspora and placing their own story in a global historical context.

On 16 February 1999, Joseph was performing seggae songs at a free open-air concert in St Louis under his stage name of Kaya.

Seggae pou liberasyon mantalite kaptivite
[Seggae is freedom from mental slavery]
Seggae pou egeye tou kalite nation ki to ete.
[Seggae brings happiness to any community you come from.]

The event had been organised to press for the decriminalisation of marijuana, and some in the crowd were openly smoking joints, police watching in the background. A stand-off lasted two days until officers moved in and made a number of arrests, including Joseph.

He was taken to the notorious Line Barracks prison,

a detention centre known locally as Alcatraz, where he confessed to smoking marijuana and awaited his fate. Much of what happened next would remain disputed, but what was beyond doubt was that, three days later, Joseph was found dead in his cell, his skull smashed and his body covered in injuries. The official record stated that the musician, suffering the delirium of drug withdrawal, had thrown himself against the walls of his cell, sustaining the fatal injuries. His friends and supporters thought a more likely explanation was that Kaya had been taken away in a police jeep to some remote site where he was savagely beaten, tortured, and most probably died. His corpse was then returned to the cell, hurled in headlong and face down, the heavy metal cell door jammed over the bare soles of his feet.

The rioting began in Roche Bois but quickly spread across the island along with news of Kaya's death. Shops, public buildings and police stations were set alight as the authorities struggled to respond. Hundreds of prisoners were liberated from the city jail, streets blocked by burning vehicles and tyres. Another Roche Bois musician, Berger Agathe, appealed for calm but was shot ninety-two times from point-blank range by police, fuelling violence which took on an increasingly ethnic character, Creole and Hindu youths fighting running battles as the black smoke rose from the rainbow nation.

A year later, an official inquiry concluded the three-day long riots had been symptoms of latent social problems.

'They represent the smouldering flames underneath the ashes that may spark off any time,' Judge Kheshoe Matadeen warned. 'The country is sitting on a powder keg.'

Without a shared sense of national belonging, the island's distinct cultural communities (*kominote*) were in competition with each other, the hierarchy of the island's history still evident in its affairs and translating into barely suppressed resentment, even hatred. The Franco-Mauritians, descendants of the plantation owners constituting just 2 per cent of the population, still controlled land, business and banking. Indo-Mauritians, representing more than two-thirds of the populace, dominated politics, police and the civil service. The Creoles, representing around a quarter of islanders, were often among the poorest and most marginalised of inhabitants. The descendants of trafficked slaves felt their ability to share in the island's economic growth was shackled by generations of injustice, resulting in what became known as '*le malaise Créole*' ('Creole ailment'). The community felt it had been largely written out of the Mauritian success story, disenfranchised and cheated.

In the years after the riots, Mauritius wrestled with the puzzle of how it might 'gather as one people, as one nation', in the words of the anthem. Communalism had created rivalry and antagonism intensified by isolation and insularity. Each community had its own narrative, all running in parallel rather than twisting together in a single national thread. It was as though the people of the island could not recognise their shared history, the collective achievement of building a prosperous settlement in the middle of the Indian Ocean from the driftwood of colonialism.

In 2009, a Truth and Justice Commission began work with a mission to repair the harms and injustices of the past 'so that Mauritius can be finally at peace with its history'. Shining a light on the dark corners of slavery was described as a form of narrative therapy, requiring, where appropriate, financial compensation and apology in order to achieve reconciliation.

The commission noted that '*métissage*' ('crossbreeding') 'is a fact of life, but little recognised or accepted' on the island. The hybridity of the Creoles, so often sneered at by the elite, was and is a key theme of the wider Mauritius story, evident among all the ethnic groups and, most obviously, the island's lingua franca. 'Mauritians have much more in common than they think they have and are much more united than they think they are,' the commission concluded.

The report urged schools to incorporate the history of all the island's communities into the curriculum. It pressed the government to create a national inventory of all the objects and places that had meaning for island communities. One specific recommendation was to create a Museum of Slavery, ideally in the old military hospital that had been built by slaves in 1740, close to the Customs House through which many of them passed on arrival.

Facing up to the truth of the past can be uncomfortable. It was telling that publication of the commission's serious and thoughtful report was barely covered by the island's newspapers and for years its recommendations were simply ignored by ministers. The 72-year-old black rights campaigner and politician Silvio Michel went on hunger strike for almost a month in 2013, demanding the government honour a commission proposal to pay compensation to descendants of slaves.

Holidaymakers arriving in Mauritius may have known little and cared less for the identity politics playing out on the island. If they expressed an interest in the island's heritage, visitors might well have been pointed to the popular Blue Penny Museum where one of the world's rarest stamps was on display, the twopence blue issued in 1847, with its profile portrait of Queen Victoria and the words 'Post Office Mauritius' along the sides. They may have been encouraged to tour the UNESCO World Heritage Site of Aapravasi Ghat (immigrant landing place), the 1849 building where many of the indentured labourers of Indian heritage were once processed. Or they may have ended up at Château de Labourdonnais, the opulent mansion built in the 1850s for the powerful sugar plantation owner Christian Wiehe. But only since the autumn of 2020 have tourists been able to visit the Intercontinental Slavery Museum in St Louis. There were 'a lot of obstacles, a lot of resistance' to the idea of a centre remembering the suffering of the island's black

Africans, a member of the museum's board of directors Jimmy Harmon explained. It took almost a decade from government approval to the centre's opening, with attempts to limit the story of the island's slave trade to a single room at the back of the building. 'I rejected that,' Jimmy said. 'No, we are not going to dilute. We're going to talk about slavery.'

The encircling shoreline of Mauritius did give the people of this beautiful island a sense of themselves. After the riots in 1999, thousands from all communities linked hands in *chaînes d'amitié* (friendship chains) that stretched from coast to coast, a powerful symbol of nationhood and islandhood. There was lots of soul-searching: Mauritian professors, preachers and poets raised inspiring voices on the importance of harmony and solidarity. The island's President set up a Committee for the Promotion of National Unity to 'instil and nurture patriotic feelings'. But the claustrophobia and vulnerability that often come with island status also excited protective instincts, encouraging tribalism and social division.

Mauritius would be cited as an example of a successful post-colonial island nation, having achieved relative economic and political stability against the odds. But beneath the veneer of the paradise island, the unhealed scars of its painful past remained.

In the past quarter of a century, billions of euros have been spent on programmes in both Northern Ireland and across the border in the Republic, looking to 'support peace and reconciliation' on the island. As part of one scheme, teenagers from disadvantaged communities on opposite sides of Belfast's peace walls are introduced to their neighbours. I watch as the two groups of young people, each indistinguishable to the untutored eye, file uncomfortably into a hall and sit at opposite ends. A jolly programme leader encourages the students into a circle for a party game and slowly one can see guards drop. Tentative conversations begin. For most of those on the course, this is the first time they have ever spoken face to face with someone from the other side.

'I was always told that Catholics were bad people compared to Protestants,' one girl tells me later. 'But not all Catholics are like that – they are actually mostly like us.'

'It's just the way they were brought up and we were brought up,' a boy agrees. 'Their families are telling them different things, but we are just all the same.'

There is almost bewilderment at how the narrative of distrust and suspicion played to them throughout their lives fails to stand up to scrutiny. Leaning against the peace wall on the Catholic side, 22-year-old Conor, a graduate of the scheme, tells me about the first time he ventured into the Protestant neighbourhood just a minute's walk away.

'It was so, so hard. Even though it is the same road, you get shivers and stuff.'

Conor remembers how he was greeted at the security gate

by Sammy, a Protestant lad he'd got to know on the course. 'That was amazing, someone just coming and saying, OK you are safe. That there was just one of the best moments for me. I'd never had a Protestant looking out for me and they actually cared about me.'

I head back to my hotel, place Pangaea back on the table and pull out a dog-eared copy of *Imagined Communities* by Benedict Anderson. The subtitle is *Reflections on the Origin and Spread of Nationalism* and, although the author writes little about the country of his blood, he has something relevant to say in terms of the rival expressions of nationalism which have divided the island of Ireland's communities. 'Regardless of the actual inequality and exploitation that may prevail in each, the nation is always conceived as a deep, horizontal comradeship,' Benedict writes. 'It is the magic of nationalism to turn chance into destiny.'

Nationhood is an artificial construct that attempts to give structure and meaning to the various traditions and cultures occupying a defined geography. An island's coastline may provide physical encouragement for the idea of shared, sovereign territory. But it is not enough. Nationhood is a sensation. The very act of drawing a boundary around a population heightens the sense of difference within and without. It creates a psychological island, an imagined community which requires mutual respect, tolerance, honesty and hard work to deal with the eternal challenges of islandness.

CHAPTER 16

PIRATES

Away from Prying Eyes

On the way from Panama to Hispaniola, during his last ill-fated voyage in search of the passage to the Indies in 1503, Christopher Columbus spotted a pair of low-lying uninhabited islands. He didn't bother stepping ashore. For his son Ferdinand, who was taking notes on the deck of Christopher's termite-infested flagship, the most notable feature of the skulking islands was the abundance of sea turtles he could see lying on the beaches and gliding in the coastal waters; greens (*Chelonia mydas*), hawksbills (*Eretmochelys imbricata*) and loggerheads (*Caretta caretta*), all in such numbers he named the place *Las Tortugas* (Turtle Islands).

Within thirty years, the reputation of the Caribbean island cluster had been revised, no longer associated with the gentle turtle but with the predatory crocodile, renamed *Caimanes* (Cayman Islands) after the Taíno word for the flesh-eating alligators that lurked in the creeks and inlets. Adventurers, privateers and bucca-neers butchered and feasted upon the turtle population around Cayman Brac and Little Cayman, as well as Grand Cayman to their west, but some unwary sailors were themselves ripped apart and consumed by the marine crocodiles (*Crocodylus acutus/Crocodylus rhombifer*) that hid in the shallows. Reptiles and pirates prowled the coves of the Cayman Islands, both waiting to pounce on the unsuspecting who ventured too close.

To this day, the exploits of the rogues who for centuries used Cayman as a criminal hideout are celebrated in a fortnight-long pirate festival upon the three islands, with mock invasions and theatrical brawls that draw tens of thousands of visitors. It is a very public show for an archipelago which sells itself to the world as a place of seclusion and discretion, a sanctuary from unwanted scrutiny.

The growing conditions of Cayman proved unsuitable for the large plantations which underpinned the slave trade, and so the islands found themselves on the periphery of Caribbean affairs. Controlled by the British from 1670, they were used principally as a fuelling stop for the supply ships servicing the sugar-cane estates of Jamaica. Merchants would navigate their way through the menacing reefs to take on fresh water and turtle meat. The rewards were minimal, certainly compared with the fortunes being made elsewhere in the West Indies. The few hundred residents included descendants of deserters and fugitives who had come to hide themselves and their booty from the public gaze, a land of lawlessness and brigandage where unscrupulous locals were active in the 'wrecking business', luring vessels onto the encircling coral and salvaging the cargo. It was regarded as a godless territory, plagued by fever-ridden mosquitoes (*Aedes aegypti*).

For centuries, Cayman struggled to shake off its crocodilian reputation, shunned as dangerous and untrustworthy. By the mid-1950s, the islands were still without running water or telephone service and only intermittent electricity. With most of the menfolk at sea on turtle-fishing boats or merchant ships, the resident population was overwhelmingly female at a time when women of the islands had yet to be given the vote. Letters from the British commissioner to the governor in neighbouring Jamaica revealed how the dependency was regarded as 'one rotten borough' where vested interests were open to bribery. The Cayman Islands had become an irksome drain on the resources of the motherland at a time when the United Kingdom was busy licking its postwar wounds and watching its empire crumble.

The Colonial Office was anxious to navigate its dependencies to a place that would no longer require substantial economic aid from London. In 1960, a former Battle of Britain fighter pilot, Jack Rose, was despatched to be the official administrator of the Cayman Islands, an imposing figure seen striding from the British West

Indian Airways turboprop shortly after landing at Grand Cayman Island Airport. Locals learned how, twenty years earlier, Jack had been shot down over the Channel only to be rescued and climb back into a cockpit a few days later. With a chestful of medals including the Distinguished Flying Cross, he had been on active service from the first day of the war until the last, defending the British coast from German air attack above the white cliffs of Dover in 1940 and commanding air support as General Bill Slim's 14th Army advanced on Rangoon in 1945.

This was the man Whitehall had selected to help the islands achieve greater self-sufficiency, a war hero with a reputation for survival and resilience. Jack surveyed the scene and assessed the Caymanian character. He concluded he had landed among a proud island people struggling with a dysfunctional economy. Their income came largely from remittances sent back from men on the turtle boats, a few customs duties and, notably, the sale of Cayman Islands stamps to dealers around the world.

Jack heard how, during the First World War, more than half the government's total revenue had come from philatelists' passion for Caymanian stamps. A local schoolteacher had suggested the islands' Post Office should raise money for the war effort by adding an extra penny-halfpenny tax to overseas mail. The overprinted Cayman Islands 'War Stamp' became wildly popular with collectors and more than a million were produced in various colours and denominations. Subsequently, an extensive range of commemorative issues were produced by the island postal service, stamps marking the most obscure of anniversaries and local events. It was, Jack must have noted, a highly creative business solution to raise desperately needed funds, although its importance underlined the unsustainability of the Cayman economy.

Tourism was growing rapidly elsewhere in the Caribbean, but the swarms of mosquitoes on the Cayman Islands were suffocating any significant growth, a pamphlet advising visitors, 'you must console yourself with the thought that life anywhere is impossible without a little healthy frustration'. If Britain was to free itself from the financial burden of this colony, Jack realised, something radical needed to happen.

His answer was to exploit the islanders' independent spirit and imaginative approach to commerce. Jack helped draft legislation that suggested a new role for the colony in financial services. Elsewhere in the Caribbean, Bermuda and the Bahamas had been registering businesses and banking their money since the 1930s, with few pesky questions or taxes. An even more light-touch model on Cayman, Jack advised local politicians, would generate fees that could be used to eradicate mosquitoes, build schools, create jobs and free the islands from the administrative orbit of neighbouring Jamaica.

Perhaps it was all those briny tales of pirates burying their treasure on remote Caribbean islands, but

Caymanians instinctively embraced the proposition. The 1960 Companies Law would allow wealthy individuals to register a business and shield their identity and their money, creating an island haven for those who sought to conduct their affairs without attracting the gimlet eye of the taxman. Modelled on English corporate law, employing similar language and structure, the legislation contained deliberate loopholes that shrewd financial lawyers in the City of London and across the English-speaking world would recognise and exploit.

As the ferry breaks free from the pull of the Emerald Isle, leaving the shelter of Belfast Lough to feel the full force of the Irish Sea, I notice a trio of low-lying islands on our starboard side. There are thousands of small islands scattered around the coast of the British Isles and these three, like many, have seen the people pack their bags and the seabirds assume ownership. Hundreds of Manx shearwaters (*Puffinus puffinus*) and Arctic terns (*Sterna paradisaea*) swoop over the Copeland Islands (*Oileáin Chóplainn*), surveying the twisted gravestones in the abandoned kirkyard on the largest of the group, a poignant relic of the community that tried to scratch a living here but which, in the end, was defeated by isolation.

There is something mesmerising about uninhabited islands, remote and wild places that hint of mysterious goings-on and dark secrets. Aware of Pangaea's sleeping form in my pocket,

I am suddenly seven years old, under the bedcovers with a torch, my private hideaway after lights out. Here I devour Enid Blyton, Arthur Ransome and Robert Louis Stevenson with their tales of old maps, salty sea captains, talking parrots and desert islands which inevitably conceal some nefarious activity. The island in these stories is an autonomous dominion, free from the restraining influences of law and order, a place where conventional rules do not apply, where exciting adventures can happen.

The wind brushes my cheeks and I am back looking at the bleak low-lying outlines of the three Copeland Islands: Big Copeland, where five or six farming families once grew oats and barley and beans in fields fertilised with seaweed, their children attending a small school; Old Lighthouse Island, where a brazier of coals was kept burning as a warning to ships before a catalogue of shipwrecks inspired the Dublin Ballast Board to spend a small fortune on a seventy-foot high lighthouse, the ruined stump now a bird observatory; Mew Island, the smallest of the three, which took over lighthouse duties in 1882 and where countless rabbits breed in the bracken (*Pteridium aquilinum*) and balsam (*Impatiens glandulifera*). Visitors are few, mostly ornithologists and naturalists who come to study this detached ecosystem. In 2013, there was excitement when a rare root-mining weevil (*Diplapion confluens*) was discovered, an agricultural pest exterminated from the rest of Northern Ireland that had found its island sanctuary.

What kind of people decided to move to these inhospitable and precarious rocks? What was the lure of the islands? Some, it would appear, were those seeking to distance themselves from social convention: a recalcitrant monk called John refused to leave Bangor Abbey's retreat on Big Copeland in the seventeenth century and stayed as a hermit, perhaps in tribute to Saint Comgall, the abbey's medieval founder, who had famously withdrawn to an island on Lough Erne; local folklore has it that the sole resident recorded living on the islands in the 1951 census was a woman who shot rabbits and ate seabird eggs to survive.

The islands, though, were also attractive to those who wished to be separated from the watchful oversight of the authorities. In the 1820s, the owner of the Copeland Islands, a local MP named David Ker, prevented the coastguard from putting a station on the archipelago. Instead of having a permanent presence, the guards were only able to visit a few times a year, their boats visible to the small population long before they arrived to do their customs checks. David's tenants were free to supplement their meagre incomes by smuggling contraband tobacco and whiskey into County Down, an illicit trade that continued for hundreds of years. The islanders were also accused of stealing the cargoes of ships that foundered on the hidden islets and skerries of Copeland Sound, with rumours that vessels were sometimes deliberately lured onto the rocks.

Islands attract non-conformists, characters who don't want to follow the crowd or the rulebook. They may be loners or dreamers, renegades or rogues, but people need a reason for seeking isolation. Humans generally look to follow social norms, the formal and informal rules which govern the way a society behaves. It is hardwired into us from birth to copy the conduct and actions of those around us. But there is an opposite force at play too, a rebellious and disobedient streak in each of us that wants to defy the norm, to demonstrate our free will and individuality. It is the contradiction that defines islandness.

The Cayman Islands had fashioned a jurisdiction that was simultaneously buccaneering and respectable, radical and safe. While other Caribbean islands were busy pursuing independence from Britain, Caymanians voted to remain a colony. Their flag incorporated the British blue ensign, their coat of arms featured a rampant British lion and their stamps portrayed the head of the British sovereign. The territory's status as a dependency of the United Kingdom bestowed an aura of propriety that gave confidence to international entrepreneurs and investors, one island attaching itself to the reputation of another thousands of miles across the ocean. Representatives of the world's super-rich began to swarm around the capital George Town, lawyers and accountants gathering like parasitic insects drawn to the scent of warm blood.

Jack was right. Foreign money washed across the beaches of the Cayman Islands in quantities locals could barely have imagined. The government was able to begin the promised mosquito eradication programme, schools were constructed, electricity and water supplies were installed. The backwater was joining the mainstream.

In 1966, the Post Office issued special stamps to mark the islands' entry into the jet age with the arrival of the first Boeing 727 at an expanded Grand Cayman airport. Another set was issued that year to commemorate the introduction of international telephone links, each stamp depicting phone lines stretching out from the islands and looping around the globe, a portrait of Queen Elizabeth II looking impressed beneath a royal crown. After centuries of being ignored, ostracised and vilified, the Cayman Islands were taking great delight in celebrating connection with the wider world.

Special stamps were also produced that year to mark the new headquarters of the World Health Organization and the twentieth anniversary of UNESCO. Caymanians were keen to attach themselves to the institutions of globalisation, but at the same time, the expat financial community was encouraging the islanders to exploit their isolation.

As the first day covers were being despatched to philatelic dealers, a raft of new legislation was passed by Cayman's assembly, laws largely drafted by canny foreign lawyers, creating an island of extraordinarily low tax and high secrecy, a place happy to put out the welcome mat for anyone with money but where the revenue inspector was strictly *persona non grata*.

Until that time, the Bahamas had been regarded as

the shadiest spot in the Caribbean to hide your loot, the islands of choice for mobsters including associates of Al Capone. But the 1967 election of Lynden Pindling, a black premier promising Bahamian independence, was too disturbing even for organised crime. Ships loaded with dollars sailed out of Nassau harbour and headed for George Town port. More cash flowed into the harbour when race riots in Bermuda led to a state of emergency that spooked investors there. The Cayman Islands, divorced from the region's post-colonial struggles and with its ties to the old white establishment in the City of London, was seen as the safest place to bury your treasure.

In London, though, the Inland Revenue and the Treasury were getting twitchy about the impact of tax havens in general and the Cayman Islands in particular. Officials were dismayed at how this British dependency was setting out, quite deliberately, to frustrate the British taxman. There was concern, too, at how sterling was disappearing into unregulated Cayman accounts, undermining exchange controls and damaging the United Kingdom's balance of payments. In 1969, the Bank of England wrote to the Treasury bemoaning the 'expatriate operators' who were turning the islands into 'their own private empires'.

But as the '60s ticked over into the '70s, the mood shifted. Some senior Whitehall voices began to whisper of the advantages of letting the islanders do their own thing. The dependency would be less dependent, it was pointed out. What was more, there were exciting opportunities for Britain's financial institutions in the rapidly expanding offshore sector. The City of London was ideally placed to offer services and expertise to businesses wishing to take advantage of the islands' distinct status. Far from being a drag on the UK economy, Cayman could prove to be a catalyst for growth, it was suggested, at a time when London's role as a global financial centre was potentially threatened by the demise of sterling as an international currency and the UK's decision to join the European Economic Community.

With Whitehall looking the other way, the Cayman Islands seized their moment, doing whatever was necessary to secure their position as the most successful 'offshore financial centre' in the world. Thousands of businesses, including hundreds of banks, registered themselves on the islands, their physical presence often no more than a brass plate on a door. By the mid-1980s, islands with around 20,000 residents were home to 20,000 companies generating a gross domestic product of more than $250 million. A sizeable chunk of the planet's money, some $127 billion in bank assets alone, was secreted behind the archipelago's almost impenetrable coral reefs. The days of needing British aid were over.

It was an extraordinary rags-to-riches story. In less than twenty-five years, Cayman had gone from being one of the world's poorest island states to one of the wealthiest. As well as financial services, tourism had boomed. Ignored by Christopher Columbus, the islands were now a fashionable destination for tens of thousands of American vacationers. The mosquitoes had been subdued, the sun was shining, the economy was thriving. And yet the crocodiles were still lurking in the swamp.

A fog is forming around the ferry, hiding and revealing as it swirls over the waves. The coast of the Irish mainland has been lost in the mist when I become aware of the shape of an island rising up ahead. It is the Isle of Man (*Ellan Vannin*) and it seems almost to be flirting with me, offering glimpses through the brume – the grey of a pebbled shoreline, the stripes of a red and white lighthouse, flashes of yellow from drifts of gorse (*Ulex europaeus*).

Pangaea shifts in my pocket as I am prompted to remember King Gorse, the revered Manxman who ruled an island empire from this place. Also known as King Orry and the King of Mann and the Isles, Godred Crovan (*Gofraid Crobán*) was a Viking mercenary who fled to Man after finding himself on the losing side at the Battle of Stamford Bridge in 1066. Within a decade, he was plotting the violent overthrow of the family which had given him sanctuary.

The history of this period is as murky as my view, but it appears that Godred turned the strategically important isle into the capital of a great dominion of insularity, a kingdom of island communities extending hundreds of kilometres north, incorporating the peoples of Bute (*Eilean Bhòid*) and Arran in the Clyde estuary, onwards to Islay (*Ìle*) and Jura (*Diùra*), Colonsay (*Colbhasa*) and Mull (*Muile*), Eigg (*Eige*) and Rùm, Skye (*An t-Eilean Sgitheanach*) and Raasay (*Ratharsair*) in the Inner Hebrides (*Na h-Eileanan a-staigh*), across the Little Minch (*an Cuan Canach*) to South and North Uist (*Uibhist a Deas/Tuath*), Harris and Lewis (*Leòdhas*) in the Outer Hebrides (*Na h-Eileanan Siar*), along with a host of smaller islands in the seas off Scotland's western coast.

Godred had founded a dynasty of the isles that remained a potent force for centuries, his descendants commanding the maritime empire from a royal palace on the semi-detached St Patrick's Isle, a tidal island on Man's western shore. Their wealth was based on a piratical tradition that could be traced back to Viking forebears, raiding ships hiding in island bays waiting for trading vessels to sail too near. The Isle of Man, standing guard at the mouth of the North Channel between Scotland and the island of Ireland, was perfectly positioned to control and prey upon the marine traffic of the Irish Sea.

Islands and pirates go together. It is partly a consequence of their strategic and geographic situation, but it is also a state of mind. Islanders tend to feel removed from the regimes and regulations across the water, protective of their autonomy and resentful of those whose proximity threatens that sovereignty. The Kingdom of the Isles harnessed these island qualities on a grand scale.

Pirates would menace the trading routes of the Irish Sea for centuries. Rory the Turbulent (*Ruairidh an Tartair*) was the feared chieftain of the MacNeil clan which terrorised shipping from their base on the island of Barra (*Barraigh*) in the seventeenth century. Neil MacLeod (*Niall Odhar*), head of the MacLeod family centred on Lewis, joined forces with the infamous pirate captain Peter Love to protect their secret island hideouts from the Gentlemen Adventurers sent by James VI to 'civilise' the Hebrides. Fittingly, Neil's last stand was upon the tiny island of Berisay (*Bearasaigh*), situated at the margin between the archipelago and the wild Atlantic Ocean in the far north of the kingdom.

The Isle of Man, meanwhile, had become a major centre for smuggling. Records suggest that around 1670, a 'company of adventurers from Liverpool' moved to the island's capital Douglas specifically to trade in contraband. French and Dutch merchants routinely used the island to smuggle their cargos of Indian cotton, silks, tea and spices into Scotland, England, Wales and Ireland, exploiting the territory's physical separation and reportedly denying the taxman the equivalent of £100 million in revenue each year. It was such a huge racket that increasingly draconian laws were passed in London to try to curb the Manx plunderers, but the rebellious island spirit endured. The Lord of Mann, the governor appointed by the English monarch, had a personal interest in the territory's economic growth, and in the centuries of back and forth between island and mainland, the islanders gradually confirmed the right to apply their own duties and taxes. The Isle of Man had achieved the island dream: a semi-detached status that granted both autonomy and protection.

The fog has lifted now, and the island's shoreline is clearer. Today, the Isle of Man is a self-governing dependency with its own Parliament and domestic law, and yet its people enjoy British citizenship, the Queen as head of state appears on all stamps and coinage and the UK government remains responsible for the island's defence.

As the ferry swings south, I can see Ramsey Bay where the ship used by the pirate broadcaster Radio Caroline was anchored in the mid-1960s, audaciously airing non-stop pop music in defiance of UK law but supported by the islanders. When the British authorities changed the legislation to cover Manx waters, the island's Parliament (Tynwald) attempted unsuccessfully to get judges to rule the decision unconstitutional. The legal wrangling, though, allowed the pirates to continue operating for a few weeks longer, frustrating the will of mainland legislators.

In 1973, when British Prime Minister Edward Heath coined the phrase 'the unacceptable face of capitalism', he was referring to the use of a Cayman-based bank account by the corrupt businessman Roland 'Tiny' Rowland to make illicit tax-free payments to a former government minister. The following year, the islands'

largest employer and most enthusiastic promoter, play-
boy Canadian banker Jean Doucet, fled from Cayman in
his private jet when an investigation found he had 'borrowed' some of his clients' gold to shore up the business.
He was later jailed.

In 1976, US authorities served a subpoena on the head
of another Cayman-based bank, Anthony Field, accusing
him of helping Americans evade taxes. A Florida court
demanded his clients' details be revealed. It was the first
direct challenge to the secrecy of the islands by a foreign
power. Expatriate lawyers and bankers huddled together
in George Town, urging the Caymanian government to
stand firm, which it duly did. Its response to the demands
of the US authorities was the Confidential Relationships
(Preservation) Law, hastily drafted legislation making
it an imprisonable offence to break or seek to break the
code of silence.

It was an audacious move, a small island taking on
the might of a continental power, almost impudent in
its rebuff to America's mighty Internal Revenue Service.
Cayman leaders, encouraged by the powerful individuals who profited from the territory's unregulated status,
had decided to act aggressively in defence of what had
become their islands' USP.

The United States was not walking away from the
fight, though. In the early 1980s, with President Ronald
Reagan demanding a war on illegal drugs, justice officials
presented evidence that up to 40 per cent of global profits

from narcotics trafficking was being covertly laundered through Caribbean islands, notably Cayman. Once again, America put pressure on Britain to put pressure on its secretive dependency to name names.

The Caymanians were being presented with a choice that went to the contradictions of islandness. Were they open or closed, connected or separate? Were they subjects of colonial masters bound by the rules of continental courts, or a sovereign people free to govern their own affairs? Did they see themselves as law-abiding subjects or spirited buccaneers? Like the pirates of old, the islanders decided they would continue to hoist both flags as circumstance required, the British blue ensign and the Jolly Roger, separately or together.

In the summer of 1984, the legislature of the Cayman Islands passed the Narcotics Drugs (Evidence) (United States of America) Law. On the face of it, the islanders had conceded to foreign demands that they open up their books and break their oath of *omertà*. But all those brackets reflected the constraints of the legislation, restricted to documents which were identified as evidence in specific drug-related crime investigations and available only to law officers in the USA who were themselves barred from sharing the contents any further.

It was a compromise that allowed the islanders to assert their propriety while continuing to protect client confidentiality, a delicate balance of public and private that would characterise their approach for years to come. Investigations into almost every major financial scandal of the next three decades would uncover the fingerprints of the Cayman Islands: BCCI, Enron, Parmalat, Citigroup, Glencore and, fatefully, the collapse of the institutions which triggered the global financial crisis of 2008, Bear Stearns and Lehman Brothers. Both banks had secreted billions in toxic assets on the islands, the stash of dodgy money that helped precipitate the tidal wave of economic misery across the planet. Throughout all of this turmoil, however, the Cayman Islands continued to claim that it was a legitimate financial centre, refusing to accept the description of their territory as a 'tax haven' at all.

For islands which had got rich by avoiding scrutiny, these were difficult times. The global crash was shining the international spotlight onto their activities like never before, its beam coming to rest on one George Town five-storey office block named after a resident billionaire, Andreas Ugland. US President Barack Obama noted how Ugland House was, on paper, home to nearly 19,000 companies. 'Either this is the largest building in the world or the largest tax scam in the world,' the President suggested. The only physical tenant on the premises, the international law firm Maples and Calder, complained there was no evidence of any illegal activity associated with the property, but the hardship caused by the economic crash meant people and politicians were looking for someone to blame.

In April 2009, the leaders of twenty of the most powerful nations in the world gathered beneath the shiny towers of the big banks in London's Docklands to survey the wreckage from the global recession. On the G20 agenda was a discussion on 'tax havens and non-cooperative jurisdictions'.

'I believe that we can, for the first time, agree the big changes necessary for coordinated action that will signal the beginning of the end for offshore tax havens and offshore centres,' the UK Prime Minister Gordon Brown proudly announced.

On the Cayman Islands, they waited anxiously to see what would happen. 'The fact is this country is one that doesn't have a lot of resources. We are just sea, sand and sun and our people,' the Cayman government's business leader Kurt Tibbetts complained to reporters ahead of the London summit. 'Without our financial sector, we'd be in much trouble to survive as a small island nation.' In fact, the Cayman Islands had become one of the richest countries on earth, with annual wealth creation equivalent to $50,000 for each and every resident.

The G20 leaders were presented with a list of tax havens and other financial centres with a rating as to how far each had complied with international tax rules. All of their own countries were given a clean bill of health and placed upon the 'white list': China, Russia, the UK and the USA all deemed honest and transparent. A 'black list' of places regarded as international troublemakers and most at risk from G20 sanctions contained Costa Rica, Labuan Island in Malaysia, the Philippines and Uruguay.

In between was a separate list of tax havens which had expressed commitment to the standards but where evaluators had decided they were 'not yet substantially implemented'. Two-thirds of the names on this 'grey list' were small islands which described themselves as offshore financial centres.

Anguilla, Antigua and Barbuda, Aruba, the Bahamas, Bermuda, the British Virgin Islands, the Cook Islands, Dominica, the Marshall Islands, Montserrat, Nauru, Netherlands Antilles, Niue, St Kitts and Nevis, St Lucia, St Vincent and the Grenadines, Samoa, Turks and Caicos, Vanuatu and, inevitably, the Cayman Islands were singled out for having failed to cooperate fully with the international tax rules.

In George Town, the news was greeted with relief. 'Only four jurisdictions have been named and shamed on the much-feared G20 summit blacklist and the Cayman Islands is not one of them,' an island newspaper delightedly informed its readers. The Cayman policy of agreeing just enough regulation to keep the big powers at bay while protecting clients from unwanted attention still appeared to be working.

'We believe that the Cayman Islands is cast in the lightest shade of grey among the financial centres with which we have been listed,' Kurt reassured a press conference.

Offshore financial centres were to come under even greater scrutiny, however, with the leaking of tens of millions of confidential financial documents in 2016 and 2017. The Panama Papers and Paradise Papers lifted the lid on who was hiding behind island secrecy laws and what they were up to. The Cayman Islands' protective reef had been breached.

It emerged that Her Majesty Queen Elizabeth II, whose head still appeared on the stamps of the Cayman Islands and who nominally appointed the island's governor, had investments in the territory. The leader of the Labour Party, Jeremy Corbyn, suggested she should apologise for keeping her money offshore. Her son and heir, Prince Charles, also had funds invested in Cayman. The journalists who broke the story were quick to point out there was no suggestion the royals had avoided paying taxes or behaved improperly, but the revelations fuelled global anger at how the world's super-rich were able to conduct their affairs without the bothersome scrutiny that applied to mere mortals.

On 1 May 2018, British patience with its capricious dependency appeared to run out. Members of the United Kingdom Parliament, backbenchers from both left and right, demanded an amendment to anti-money laundering legislation that would oblige all British Overseas Territories to publish the names of everyone who owned a company registered on their islands.

Conservative MP Andrew Mitchell urged the

government to bear down on corruption, tax evasion, terrorist financing and fraud. 'Much of the money, as the Paradise Papers and the Panama Papers make clear, passes through British Overseas Territories.'

The Labour MP Dame Margaret Hodge rose to speak. 'We cannot sit here and ignore the practices that allow Britain and our British Overseas Territories to provide safe havens for dirty money,' she said. 'If we can act to root out the corruption, we must do so. Our proposal is simple but powerful. It is easy to implement but lethal in its effectiveness.'

The UK government had not agreed or expected the move, but such was the feeling on the backbenches that ministers feared a defeat and agreed to waive the amendment through.

News of the surprise vote was greeted with fury on the Cayman Islands. Premier Alden McLaughlin described it as 'colonial despotism' and said the islanders would simply not accept it. 'It is not just the future of our financial industry that is at risk, it is our very existence,' he said angrily. Caymanians had been proud to use their British identity to bolster a reputation for respectable and lucrative business, but now their attachment to the blue ensign was threatening their way of life.

Were the islands about to break their ties with the motherland, to pursue a new role as an independent nation? There was fevered discussion among the political elite as they weighed up their options. Once again, the Cayman Islands were being asked to address a fundamental of islandness, their relationship to the world beyond the white coral sand.

A fortnight later, Alden emerged from his office to brief reporters. 'I don't think we are at this stage ready, as a country, as a people, as a government, for independence,' he told them. 'I personally am concerned about the viability of a country of 65,000 people as an independent nation in this big, bad world.' The rhetoric painted a picture of a small, vulnerable island at the mercy of great continental powers, a scapegoat for other people's economic mismanagement and the victim of libellous slurs. Cayman reluctantly agreed to publish a public registry of business owners. However, the list would not be produced in 2020, as the UK Parliament had demanded, but by 2023. Cayman also drafted legislation which would oblige registered companies to have an adequate physical presence on the islands, potentially an end to the 'brass plate' activities of buildings like Ugland House that had so enraged President Barack Obama.

It was not enough. In February 2020, with a virus about to engulf the world's economy, the European Union announced that the Cayman Islands were now to be placed upon the 'much-feared' blacklist, deemed a non-cooperative jurisdiction for failing to implement promised reforms in time.

Pirate Week was cancelled that year because of the pandemic. The British governor was not kidnapped and paraded through the streets by cutlass-wielding brigands as tradition had come to dictate. No one was presented with the black spot, that famous act of piratical judgement invented by Robert Louis Stevenson for his novel *Treasure Island*. But Caymanians may have felt their entire island had been shown the black spot as they continued to protest their innocence, vowing to convince the world of their righteousness.

The sun still shines on the Cayman Islands, the Caribbean Sea still laps upon their shores, turtles still graze in the coral and crocodiles still wait in the shallows.

In the second half of the twentieth century, many islands finally emerging from the depths of colonial oppression were searching for their identity and purpose. All bore the scars of their history, the weals and welts inflicted by subjugation and social engineering. But releasing the chains also exposed the vulnerabilities of island status, the trials of isolation. As the community of the Copeland Islands had discovered, living independently on a remote island can be an ordeal better left to the birds. It is, though, a challenge people have been ready to take on because, instinctively, we are encouraged to test our resilience in the search for personal freedom. The thrill of those island stories I read by torchlight under the covers of my bed comes with the separation from the security and rules of the mainland. Psychologically, islands encourage adventurers and outlaws, and that helps explain why so many island states have chosen to risk the wrath of their old masters by becoming havens for secrets.

Progress is led by the non-conformists, the individuals who think and act differently from the herd, who refuse to be shackled by convention or custom. Creativity is the antonym of orthodoxy. It is about breaking the rules in pursuit of the new and better, and so evolution has retained a thread within our fabric that dares us to challenge authority and celebrate our uniqueness. There is a pirate in all of us. But the freebooting rebel must be balanced with a compliant citizenry or societies cannot function. Without laws there cannot be order and without order there cannot be peace and without peace there cannot be well-being. The paradox of the human condition is found in the same place as the paradox of islandness, upon the shore where 'us' becomes 'them'.

CHAPTER 17

FLAGS

Staking a Claim

I place Pangaea on the bookshelf, absentmindedly running a finger along my career. Lined up in date order in my study are jotters and notepads that go right back to my first days as a cub reporter on the local paper, a chaotic archive stuffed with cuttings, business cards and page upon page of almost illegible shorthand. At various times, I have tried to impose organisation upon this scribbled muddle, imagining there is some important truth to be found within decades of journalistic musing and doodling. On the spines of some notebooks is the evidence of date labels, most now faded, peeling off or missing entirely.

With Pangaea feigning disinterest, I pull out a volume identified as 'Mar–Apr 1982'. I was then working as a producer on the London radio station LBC, putting together the evening news programme. Flicking through the notepad, my eye is drawn to a page with 'Monday 22 March' scrawled at the top. Beneath is a list of possible stories for the programme, with a tick, question mark or cross beside each item. I am struck by two entries on the prospects list. One story is slugged 'Fossil' and another 'Scrap Metal'. Deciphering my shorthand notes, I realise a strange coincidence: both stories are about discoveries by scientists upon the same remote spiral of islands which twists anti-clockwise from the Antarctic Peninsula and into the South Atlantic Ocean. 'Fossil' is marked by a single cross. 'Scrap Metal' has a question mark which I have cancelled out and then another cross. Neither story made the cut that evening.

I sit down to remind myself of the details. 'Fossil' relates to a briefing given by an American palaeontologist, William J. Zinsmeister. He had just returned from Seymour Island, a deserted fragment at the north-eastern tip of Antarctica, territory claimed by several nations including the United Kingdom. To William, however, Seymour belonged to its extraordinary fossils. The island had been nicknamed the Rosetta Stone of Antarctic palaeontology.

For three weeks, William had waded through thick mud and endured gale-force winds on Seymour, searching for what some considered to be the most sought-after prize in geology. Just as he and his team were preparing to pack away their rock hammers and concede defeat, they found it. From the sludge, a researcher identified jaw pieces and teeth from a small berry-eating marsupial (*Polydolopida*), a long-extinct forest-dwelling creature about the size of a squirrel. Their excitement was almost uncontainable. This inhospitable island had delivered up the key to how evolution and Alfred

Wegener's continental drift had combined to place marsupials in both South America and Australia, an ocean apart. The fossil was hailed as 'one of the most significant discoveries' of recent times.

From that day, scientists could show how ancestors of South America's pouched opossums had migrated across the southern half of the Pangaean supercontinent some 50 million years ago. They had travelled through thick forests of towering trees to reach territory that was breaking away to form the Australian continent – one of Pangaea's children. Here, they would evolve to become kangaroos and wallabies (*Macropodiformes*), wombats and koalas (*Vombatiformes*), bandicoots and bilbies (*Peramelemorphia*), possums and gliders (*Phalangeriformes*), a menagerie of more than 150 marsupial species found nowhere else on the planet. Seymour Island, a disputed and isolated speck cast adrift from the protective icesheet of Antarctica, had yielded the answer to one of the natural world's most intriguing questions.

The second story, bearing the strapline 'Scrap Metal', was from another contested island territory in the same volcanic arc: South Georgia. This was, I recall thinking at the time, an odd and rather troubling tale. Four members of the British Antarctic Survey, UK scientists based on the island, had gone to an abandoned whaling station at Port Leith after seeing a strange ship in the harbour and hearing shots. They arrived to find a reindeer being barbecued over a fire, a table set for twelve and a group of men, some wearing white paramilitary uniforms. The black and scarlet sign on the derelict building had originally read 'British Antarctic Survey / Unauthorised Entry Prohibited / *Entrada Prohibida Sin Autorisacion*'. Now, the word 'British' had been replaced with the word 'Argentine'. The scientists also noticed, with a shudder, that the flag of Argentina had been hoisted upon a nearby tower.

The men claimed to be scrap metal merchants who had come to recycle material from the disused whaling base, but the strange scene suggested their intentions might be rather more sinister. Subsequent events would prove that indeed they were.

I regret that neither 'Fossil' nor 'Scrap Metal' found its way from my prospects list to the radio that night. I suspect that both island tales had felt too obscure for the drive-time audience, too distant from the concerns of those negotiating London's evening rush hour.

The programme, my notes tell me, was instead dominated by 'Greenham', the news that around 250 women had blocked access to the Royal Air Force base at Greenham Common in Berkshire, with police making dozens of arrests. The women had been protesting against the UK government's decision to allow the United States to move some of its arsenal of nuclear weapons to the airbase in southern England. The Cold War had become distinctly icy and positioning strategic cruise missiles (BGM-109G Gryphon) on the island of Great Britain would put the Soviet Union in direct range of America's nuclear warheads. The women opposed nuclear weapons in principle, but they also argued that the US was treating the UK as if it was the fifty-first state, disregarding the autonomy of their precious island territory.

With Pangaea watching me, I reflect on how, in different ways, the three stories that had jostled for their place in my running order that day back in 1982 had all been about the occupation of island space. Dominion over territory within the shoreline of an island is generally seen as a product of its backstory, the relationship between the land and the living things that physically occupied the land across time. The longer and more intense that association, the greater the claim.

'Fossil' was about a long-extinct creature which had physically occupied the soil of Antarctica for 50 million years, an island tale of both transience and permanence. 'Greenham', I realise, was a story of women asserting their sovereign rights over a foreign power by ritually occupying island territory. 'Scrap Metal', although I did not know it then, was the prologue to the story of another woman who would seek to assert sovereign rights over island territory by physical occupation of the space.

The story of islands is often a tale of shifting sovereignty. To those living within the contours of a shoreline, the territory has a direct link to their identity and way of life. To those aboard hostile ships coming over the horizon, the island is a strategic prize to be won. If control changes hands, and history is littered with islands which have experienced this multiple times, then the relationship with the land also switches.

I flick forward a few pages in my notebook. The entry is headed Saturday 3 April, a day I wouldn't normally have been on shift. But we were mustered in the newsroom because, as the Prime Minister had explained to a hastily recalled

Parliament that afternoon, 'for the first time for many years, British sovereign territory has been invaded by a foreign power'.

Pangaea appears to be watching me as I decipher my shorthand of Margaret Thatcher's statement to the House of Commons. The Falkland Islands, 500 kilometres from the South American coast in the South Atlantic, with a population of fewer than 2,000 souls, had been contested territory for centuries. Now, the Prime Minister told a shocked chamber, significant Argentine forces had crossed its beaches and the British governor had been left with no option but to surrender.

To Argentina, they were *Islas Malvinas*, an archipelago strewn on the Patagonian Shelf, connected by eternal geology and geography to the tip of South America and the Argentine state. To the United Kingdom, the Falklands and their neighbour South Georgia were 'British Overseas Territories', umbilically attached to the United Kingdom by centuries of history and custom.

I go back to my notes from that day. 'The people of the Falkland Islands, like the people of the United Kingdom, are an island race ... British in stock and tradition ... with the right to choose their own way of life and to determine their own allegiance.' Margaret was summoning shared islander identity to press her point, her message that the hostile intentions of a continental foe would be met by the pluck and resilience of island people. Once the national story had been of Britain as the great imperial force straddling the great continents of the globe. Now it had been rewritten as the tale of a small island resolute in the face of danger, of shorelines protected and defended with the determination and courage of an underdog.

'They are few in number, but they have the right to live in peace,' the Prime Minister said of the Falklanders. Inside and beyond the walls of the Palace of Westminster, the phrase evoked memories of 'the few' in Spitfires and Hurricanes over white cliffs, the brave airmen protecting island sovereignty against invasion in 1940. 'Their way of life is British; their allegiance is to the Crown. It is the wish of the British people and the duty of Her Majesty's government to do everything that we can to uphold that right,' my scrawled shorthand reads.

As the drumbeats of war became louder back in the spring of 1982, I recall researching the story of the islands, trying to understand their significance. The Falklands appeared to be inconsequential and inhospitable volcanic outcrops in the middle of nowhere, but their influence clearly extended far beyond their shoreline, physically and symbolically linked to national pride and global ambition. In a way, the struggle that was about to ensue turned the concept of islandness upon its head: it was not about isolation and separation but connection and control.

In the middle of the Indian Ocean, on a coral atoll wreathed in coconut palms and perfect white sands, there once lived a people who simply called themselves Îlois (the islanders). It was their birthright and their downfall to inhabit one of the most remote island territories on earth.

A nine-day sail from the tip of India and more than two weeks from the coast of Africa, the isolated tropical archipelago was once given the name Fōlhavahi or Hollhavai by Maldivian mariners, their ancient folklore telling the stories of numerous castaways marooned upon these quintessential desert islands. Portuguese explorers christened the treacherous coral banks Bassas de Chagas, the wounds of Christ on the cross, a miserable appellation that later softened to become the Chagos Islands.

The first settlers probably arrived around 1769 when French sailors planted their royal standard on the beach of the largest island in the chain, Diego Garcia, claiming the territory for King Louis XV. Licences were issued for coconut oil and copra plantations, with enslaved labour imported from Madagascar and Mozambique. France's flag was briefly replaced by Britain's Union Jack in 1786, before French sailors tore it down and restored their claim a few months later. After Napoléon's abdication in 1814, however, the colours of George III were once again raised over the territory and the residents became subjects of the British Crown.

In 1957, the Colonial Film Unit was sent to record life in this remote and rarely visited corner of the empire. They found small villages of thatched cottages, a school busy with white-shirted children and a church with a graveyard marking the resting places of ancestors stretching back five generations. The island swayed to the relaxed rhythm of a contented community numbering around 2,000 British citizens, many of them black descendants of workers introduced by plantation owners almost two centuries earlier. What the film crew did not document, however, was the arrival on Diego Garcia that year of Admiral Jerauld Wright, commander of the US Atlantic Fleet. He was the first in a succession of American naval leaders who crossed the white beaches to cast a cool strategic eye upon the atoll.

With the Cold War intensifying and Soviet naval operations expanding into the Indian Ocean, the Pentagon had been on the hunt for somewhere from which it could patrol the region. At a briefing in 1964, after various alternatives had fallen through, the vice-chief of US naval operations, Admiral Horacio Rivero, pointed to Diego Garcia on his map and announced, 'I want this island!' Officials briefed him on what was called 'the problem of the Îlois', but the military was insistent the island must be 'swept and sanitised' of any local population.

The British were keen to assist their American allies, not least because they wanted the US to offer a substantial discount on the purchase of Polaris missiles, the business end of an independent nuclear deterrent that helped the UK retain international clout as its empire declined. This prize, however, required the removal of the community on the Chagos Islands, and so began a truly shameful episode in recent British colonial history.

The first step in the United Kingdom's plot to disown its own citizens on the Chagos Islands was to uncouple the administrative connection between the archipelago and Mauritius, a British colony that was inching towards independence. In 1965, the Mauritian government was persuaded to give up its links to Diego Garcia and the surrounding islands in return for £3 million, at which point the area was renamed British Indian Ocean Territory (BIOT). The Foreign Office then hatched a secret plan to disenfranchise the Îlois residents by claiming they were transient workers, rather than a community of British subjects whose occupation of the island stretched back generations.

The most senior official at the Foreign Office, Sir Paul Gore-Booth, explained the strategy in a confidential memorandum. 'We must surely be very tough about this. The object of the exercise was to get some rocks that will remain ours. There will be no indigenous population except seagulls.'

A handwritten reply from Denis Greenhill, head of the UK's Joint Intelligence Committee, was even more shocking when it came to official attitudes towards the black British population of the Chagos Islands. 'Unfortunately, along with the birds go some few Tarzans or Man Fridays whose origins are obscure, and who are hopefully being wished on to Mauritius etc. When this has been done, I agree we must be very tough and a submission is being done accordingly.'

The Foreign Office decided to send Sir Bruce Great-batch to do the job, a diplomat who had worked extensively in Africa. His depopulation strategy was brutal. Any islander who needed to leave the territory for medical treatment was barred from returning. The coconut plantations were gradually shut down, throwing hundreds of people out of work. The ship that brought the post and basic food supplies was prevented from docking, effectively blockading the islands.

Desperate islanders were told they must leave their homes with only a single suitcase of belongings, before being loaded onto ships, dumped on outlying islands or taken directly to Mauritius. Here they were forced to live in abandoned slums without running water or electricity. There was no compensation and little help with resettlement. On one stormy voyage, families were given a single straw mattress and placed in the hold of a cargo ship alongside a consignment of seabird guano.

Sir Bruce regarded the Îlois as 'unsophisticated and untrainable' and his treatment of them was profoundly inhumane. In the early 1970s, with US naval battalions already constructing a vast runway for their bombers and a satellite spy station to maintain a watchful eye on the region, he came up with a new tactic to intimidate the local people still clinging on to their homes and way of life. He ordered the private management agents working for the BIOT administration on Diego Garcia, Moulinie & Co., to kill all the islanders' pet dogs. The agents first tried to shoot and poison the animals, but this proved too slow. They then used raw meat to lure around 1,600 dogs into a sealed copra-drying shed where they were gassed to death using the exhausts from two American military jeeps. Once the howling had stopped, the building was set alight, and the dogs' carcasses burned to ashes in front of screaming Îlois children and their families.

Pangaea keeps a sleepy eye on me as I dig around in the ice and mud of the Falkland Islands' backstory. The first recorded landing was by the English mariner John Strong in 1690. He named the strait between the two main islands Falkland Sound, honouring the Viscount of Falkland who had sponsored his expedition and initiating a naming trend that continues in the region to this day.

Ships flying various flags came and went, French and Spanish and British, with a few dozen people stationed on the islands from time to time to demonstrate a colonial presence. The archipelago was not an enticing place to settle permanently and was used primarily as a military outpost and penal colony. Various gauchos, whalers and sealers did try to scratch a living from the Falklands, but it was hard going. In 1774, the British decided to withdraw their personnel, taking the precaution of erecting a plaque stating that the Falklands belonged to King George III. In their absence, however, various claims were made upon the islands' sovereignty. Argentinians occupied the archipelago for a few years, a situation which prompted the British to send their garrison back to the territory, formally declaring the Falklands a Crown Colony in 1840.

The following year, the Colonial Office in London appointed a governor to the islands, Richard Moody, just twenty-eight years old but with an impressive beard and military pedigree. Like his father before him, Richard desired nothing more than to paint the world British pink and the Falklands posting was a chance personally to wield a brush. He stepped across the shoreline of the islands and immediately set about surveying the territory. His assessment was that establishing a stable British colony on the islands required two ingredients: islanders and sheep. 'The settlers best adapted to colonise these islands would be from among the industrious population of the Orkneys and the Shetlands,' he told London. They were 'accustomed to a hardy life and as much seamen as landmen', he explained. Richard's 'General Report' also recommended the importation of thousands of Cheviot sheep to give purpose to the colonisation.

At that time, physical possession or 'effective occupation' of land was increasingly recognised internationally as an important element of sovereignty: flying your flag, establishing an administration, maintaining order and running an economy all gave weight to territorial claim. In the context of desolate islands, presence was a decisive factor. Hundreds of Scottish islanders were encouraged to make the journey to the South

Atlantic and establish farms on the treeless and windswept Falklands. The returns were meagre, the life was hard, but the colonists came with a patriotic purpose.

Sitting at the northern end of the volcanic coil that draws a question mark in the South Atlantic, the Falklands are on the edge of habitability. As one travels around the curl of islands – to South Georgia and South Sandwich Islands to the east, down to the South Orkney Islands, across to the South Shetland Islands and then along the twisting Antarctic Peninsula – sustained human occupation becomes increasingly unrealistic. The Union Jack flying over the governor's office in the Falklands capital Port Stanley was used to justify a territorial claim that incorporated thousands of largely deserted islands and a sizeable wedge of continental Antarctica, culminating in a point at the South Pole itself.

I unroll a map and Pangaea helps hold it flat on my desk. 'Chart of the World', it proclaims, 'Showing the Extent and Distribution of the British Empire'. Produced in 1909, the map glows with imperial pink, distributed on every continent and linked by 'Great Lines of International Commerce', which bounce between countless pink islands in every ocean. Towards the bottom of the chart, I can see the Falkland Islands connected to this web of commercial power with a line that sweeps around Cape Horn to New Zealand and Australia, from where it drives up to Ceylon (Sri Lanka), on to the Seychelles and Mauritius, around the Cape of Good Hope and north to the islands of St Helena and Ascension before arcing up into the English Channel and finally to London. Right at the foot of the map, in the emptiness of Antarctica, I notice the cartographer has painted a portion bright pink. The British Empire, the document declares, extends to every corner of the globe.

Pangaea appears to shuffle on her low couch as I continue to explore the stories of these faraway islands, distant and remote places on the edge of the map and most people's consciousness.

In 1942, with the Second World War at its height, a report arrived in London warning that a counterclaimant for sovereignty of the Antarctic islands was emerging from hibernation. The Foreign Office cable informed ministers that Captain Alberto J. Oddera of the Argentinian Navy had recently guided his frigate, *1º de Mayo*, into the harbour of Deception Island, a horseshoe-shaped volcanic outcrop claimed by the United Kingdom as part of the South Shetlands. A party had gone ashore at Whalers Bay, where a large Argentine flag was painted on a disused fuel tank of the defunct Hektor Whaling Company factory. Alberto had then conducted a ceremony, placing a cylindrical canister upon the grey sand containing a certificate formally claiming sovereignty of the island on behalf of the government in Buenos Aires. The vessel sailed on, with Alberto raising flags and placing similar canisters on two more islands closer to the Antarctic Peninsula, a rock in the Melchior archipelago he named *1º de Mayo* and Winter Island further south.

The message set alarm bells ringing in the offices of Whitehall. On 28 January 1943, Britain's Foreign Secretary Sir Anthony Eden presented a confidential paper to the War Cabinet in the Prime Minister's room in the House of Commons revealing proposals for a secret mission to Antarctica. Ostensibly, the expedition (later codenamed Operation Tabarin after a Paris cabaret club) was to counter the threat from German U-boats on allied shipping in the South Atlantic, but in fact there was no significant danger. The true purpose was a response to the activities of Alberto Oddera of the Argentine Navy.

Sir Anthony explained how he had instructed the British ship HMS *Carnarvon Castle* to retrace the route of the 1º de *Mayo*. On arrival at Deception Island earlier that month, the crew had been ordered to destroy all evidence of Argentinian presence, replacing the blue and white triband with the Union Jack. Signs declaring the islands British Crown Land had also been erected, but Sir Anthony suspected the notices would not be enough to prevent further incursions. (He was proved right when, a few weeks later, Argentine ships returned to Deception and removed the British signage, once again painting the Argentine flag on the crumbling fuel tank.)

Sir Anthony told the War Cabinet that the islands of Antarctica were rightfully British. 'We hold a title by discovery,' the Foreign Secretary explained, 'although we have never been in effective occupation of them,' he conceded. The Secretary of State for the Colonies, Oliver Stanley, urged action, warning that the British Empire would be weakened 'if we allow ourselves to give up territories through indifference'. The First Lord of the Admiralty, Albert Alexander, agreed, suggesting the islands could be strategically valuable 'if the route round Cape Horn becomes more important'.

The War Cabinet gave Operation Tabarin the go-ahead and preparations began. The ambitious mission was to set up permanent British bases in Antarctica, a physical presence that would be retained throughout the polar winter in the belief that such a commitment would lend significant weight to the claim of sovereignty. While living in the most hostile of conditions, the expeditionary team would conduct scientific surveys of the area: among those recruited for the trip were a zoologist, a biologist, a geologist, a meteorologist and even a lichenologist to help record the tiniest detail of this largely unexplored region.

The fourteen-man party boarded the polar research vessel RRS *William Scoresby* in the Falklands and sailed south at the end of January 1943. The first stop was Deception Island, where five of the team went nervously ashore, wondering if they might encounter an Argentinian military force. To their relief, the only sign of their presence was a painted flag which they erased as the old whaling factory was repurposed as a research station. The remainder of the party set up another base at Port Lockroy on tiny Goudier Island, just off the Antarctic Peninsula.

The Colonial Office had been debating how else it might promote the United Kingdom's territorial claims in the region and had come up with a very British solution: stamps. Sheets of ninepence blue Falkland stamps, depicting King George VI beside a picture of the RRS *William Scoresby*, had been delivered to the team on Goudier, and on 23 March, the Port Lockroy Post Office began operation. Radio operator Fram Farrington was appointed postmaster, supervising the overprinting of the stamps with 'Graham Land Dependency of' above the words 'Falkland Islands'. 'Graham Land' was the British name for the Antarctic Peninsula, named in honour of an admiral. (In Argentina, it was *Tierra de San Martín*, venerating one of the country's founding fathers; to Chileans, it was *Tierra de O'Higgins*, after one of their founding fathers; Americans called it 'Palmer Peninsula' after a US sea captain.) The simple red lettering on the stamps was potent propaganda. Surrounded by hundreds of oblivious gentoo penguins (*Pygoscelis papua*), the concept of British territorial sovereignty extending thousands of square kilometres from Port Stanley to the South Pole had been given a measure of philatelic credence in a small hut in Antarctica.

In 1945, to consolidate British authority over the ice and rock of the southern continent, the Foreign Office established an Antarctic Place-Names Committee, which claimed responsibility for naming the features of the territory. Flett Buttress, Donnachie Cliff and Blyth Spur honoured members of Operation Tabarin, and many other parts of the landscape were named after British scientists and servicemen who had braved the Antarctic winter. To this day, no Argentinian or Chilean has been honoured by the committee.

In 1946, the tiny Post Office in Port Lockroy issued the first 'Falkland Island Dependencies' stamp, a halfpenny green with a portrait of the king adorning a map of the region. A wedge of sovereign British territory is identified by a dotted line, within which the artist has marked 'Falkland Is, South Georgia, S. Sandwich Is, S. Orkney Is, S. Shetland Is, Graham Land and South Pole'.

By 1973, the entire community of British citizens on the Chagos Islands had been removed, the territory now effectively controlled by the military forces of a foreign power. For the US and the UK, the value of the archipelago was solely strategic, exclusive occupation the mechanism for gaining tactical geopolitical advantage. For the Îlois, however, the last passenger boat steaming out from Diego Garcia marked the start of a long and continuing journey to return to their homeland, a struggle for justice that united island peoples around the world against imperial and continental might.

Mauritius, a former island colony, saw the grim treatment of its Chagossian community as symbolic of a pervasive disrespect for islanders' rights and resolved to take the battle to the international courts.

An early challenge came in 2010, when the UK government unilaterally declared the establishment of the Chagos Marine Protected Area (CMPA), the largest marine reserve in the world, encircling all the coral islands of the archipelago. Ostensibly, the huge exclusion zone covering

250,000 square miles of ocean was designed to protect the unique ecosystem of the reefs. But a leaked US State Department cable suggested another motive. Restricting fishing in the area was seen as the 'most effective long-term way to prevent any of the Chagos Islands' former inhabitants or their descendants from resettling'. The same cable also made it clear that marine protection measures would place 'no constraints on military operations'. Personnel from the US military base were free to pollute and fish the waters as they wished.

The Mauritian government challenged the establishment of the marine protection area at the Permanent Court of Arbitration in The Hague. In 2015, the court ruled that the creation of the CMPA was illegal under international law, a final and binding decision that the UK refused to respect.

Mauritius also pursued a case against the British for the 'ethnic cleansing' of islands it claimed were still part of its territory. In 2019, the United Nations General Assembly passed a resolution affirming that the Chagos archipelago 'forms an integral part of the territory of Mauritius' and demanding the UK 'withdraw its colonial administration unconditionally within a period of no more than six months'. Despite having only five nations out of 178 supporting it in the vote, Britain ignored the ruling and continued to claim there was no doubt over its sovereignty of the territory.

Taking a leaf out of the UK's book on philatelic

propaganda, Mauritius issued a series of stamps to press its argument. A 10-rupee stamp featured the words 'Republic of Mauritius – Chagos Archipelago' above a map of the islands. Two more, the 13 rupees and 25 rupees, featured the atolls and a picture of the International Court of Justice, the ICJ having advised that Britain should relinquish the territory, the US base should be dismantled, and the islanders be allowed to return home.

Emboldened by the ruling of the UN courts, on Valentine's Day 2022, a small group of Chagossians sailed back to their beloved archipelago. Stepping onto the white beach of Peros Banhos, each islander knelt and kissed the sand. A flagpole was erected and the Mauritian standard hoisted as the country's anthem was sung.

'We are performing the symbolic act of raising the flag as the British have done so many times to establish colonies,' the Mauritian ambassador to the UN, Jagdish Dharamchand Koonjul, said. 'We, however, are reclaiming what has always been our own.'

'I left my beautiful island when I was four years old,' Olivier Bancoult told the handful of journalists who had joined the trip. 'We used to live as one large family. It was paradise. My dream is to settle in my birthplace.'

Olivier laid flowers in the abandoned church where he had been baptised more than fifty years earlier, before joining other islanders to dance on the beach.

CHAPTER 18

TERRITORY

A State of Belonging

I reach up to my shelf of old notebooks again and pull out a battered volume dated 'December 1988'. I want to revisit my notes from the time I visited the Falkland Islands for the BBC. I am puzzled by the phrase 'slab of beer' which I have scrawled in the margin, until I remember how we had landed at the bleak military airfield on Ascension Island on our way south where I watched squaddies attempt to consume twenty-four cans of ale during the short stopover. They were not allowed to board the Lockheed TriStar transport plane drunk, and they couldn't drink while on board, but the rapid consumption of the 'slab' in the middle of the Atlantic allowed them to reboard the aircraft before the alcohol kicked in and the rest of the eighteen-hour flight could be experienced in a stupor.

My notebook indicates how I used the time on the plane to review cuttings from the Falklands War, reminding myself how the Argentine invasion had been felt as a body blow to national pride. 'IT'S WAR', screamed the front page of the *Sun* newspaper as the Ministry of Defence hastily assembled a naval task force to defend the British colony. The aircraft carrier HMS *Invincible* left Portsmouth on 5 April, large crowds gathering on the harbour's Tudor fortifications to wave at the sailors lined up upon deck, as brass bands played patriotic tunes. A small wooden boat bobbed precariously in the Solent, its occupants gamely holding up a union flag, a metaphor for the makeshift nature of the military response – mothballed warships hurriedly brought back into service, cruise liners requisitioned as transport vessels, car ferries repurposed as troop carriers. The spirit of Dunkirk had been revived as the rather unusual fleet set sail, the mission portrayed on the front pages as a proud island nation once again attempting to defy the odds against a tyrannical continental foe.

The TriStar vibrated noisily as squaddies snored, cruising high above the route taken by the naval task force seven years earlier. I pulled out another cutting from *The Sun* with the headline 'INVASION', reminding me how, as the improvised armada continued its three-week 13,000-kilometre voyage to the South Atlantic, a small contingent of Royal Marines had successfully retaken the island of South Georgia. The British signal read: 'Be pleased to inform Her Majesty that the White Ensign flies alongside the Union Jack in South Georgia'. The Argentine flag had been torn down. 'Just rejoice at that news,' Margaret Thatcher had told the nation, aware of gloomy assessments for achieving victory against General Leopoldo Galtieri's forces embedded on the Falklands.

The TriStar complained as the captain began its descent. Before touching down, I tried to memorise the grim numbers behind the Falklands War: a conflict that lasted seventy-four days and saw the deaths of 649 Argentine military personnel, 255 British military personnel and three Falkland Islanders.

'VICTORY', *The Sun* had proclaimed, as a photograph was sent around the world showing Royal Marines raising the Union Jack over Port Stanley once again. I remember studying the black and white image just as the plane came into land. It was a far cry from the theatrically staged image of US soldiers raising the Stars and Stripes on the island of Iwo Jima in 1945. Slightly out of focus and on an angle, the picture had an honest spontaneity about it. I also noticed that the flag being pulled up the pole was not, as the headlines suggested, the familiar Union Jack, but the flag of the Falkland's governor with the island's coat of arms in the centre. The question this image posed in my mind was whether the soldiers were genuinely reclaiming the territory for the islanders, or if the military campaign had more to do with strategic advantage for the UK government.

'Whoever controls the Indian Ocean will dominate Asia. This ocean will be the key to the seven seas in the twenty-first century. The destiny of the world will be decided on its waters.' These were the words of the celebrated US naval strategist Alfred Mahan, crystal-ball gazing back in

1897. More than a century later and the evidence was that his prediction was spot on. The Indian Ocean's sea lanes had become the main arteries of globalisation, channels for the oil and commodities that power the planet's economy, marine motorways plied by mega-ships, each capable of carrying tens of thousands of tons of cargo.

Superpowers were said to be playing 'The Great Game' in the Indian Ocean, a reference to the nineteenth-century jockeying between British and Russian empires over control of central and south Asia. The 21st-century version saw China and India going toe to toe for influence and control in the mighty and largely uninhabited waters that stretch from Africa to Australia. The scattering of remote coral reefs that once terrified mariners became strategic hubs, vital ports at the crossroads of international commerce. It was small wonder that Britain and America were so reluctant to relinquish their occupation of Diego Garcia in the Chagos Islands, the western powers keeping watch on proceedings from US Camp Thunder Cove.

Five hundred miles to the north of the base lay the Maldives, an Islamic republic made up of more than a thousand islands arrayed across the ocean in a double chain of twenty-six coral atolls. To many in the west, this was the exotic location for a luxurious escape from the stresses of normal life, but to the Chinese, it was a key piece in their strategy for global commercial domination. The idea of using islands to create geographical power lines had a long history in China, and their latest plan was to assemble a 'String of Pearls', a network of military and commercial facilities extending from the Chinese mainland to the Horn of Africa. Beijing bought control of ports on the islands of Singapore, Java, Sumatra and Sri Lanka. Where no suitable natural islands existed, China simply created them, building artificial islands in the South China Sea and the Strait of Malacca. Also among the pearls upon its string of islands were the Maldives.

In 2014, Xi Jinping became the first ever Chinese head of state to make an official visit to the island state. In the year after his trip, China lent more than a billion dollars to the Maldives, funding the construction of 11,000 apartments on an artificial island, extending the electricity grid and expanding the international airport. Tourists heading off to one of the many island resorts scratched their heads at huge screens around the building site adorned with communist slogans: Unity and Struggle; Executing Orders; Seeking Truth.

The loans also helped pay for the mile-long China–Maldives Friendship Bridge connecting the Maldivian capital Malé to the airport, the first physical link ever made between the country's islands. During construction, Chinese and Maldivian flags flew side by side as the development reached across the lagoon. There was regional unease at the way this relatively poor island state was becoming saddled with high-interest debts that it could never hope to repay, concern which turned to anger in some quarters when three Chinese warships docked in Malé harbour on a 2017 'goodwill visit'. Beijing insisted that the loans had been provided 'in accordance with the wishes and development needs of the Maldives', but to many it appeared that China was undermining the independence of the islands for its own strategic advantage.

The Indian government, which historically had enjoyed close relations with the Maldives, decided it needed to respond. A surprise change of government in Malé in 2018 was the spur for India's Prime Minister Narendra Modi to make a series of visits to the Maldives, his officials agreeing to fund the construction of another bridge, three times longer than the Friendship Bridge. The project, costing half a billion dollars in grants and loans, connected Malé with the neighbouring islands of Villingili, Gulhifalhu and Thilafushi. Indian officials told reporters that the new bridge, constructed with New Delhi money, would render its Beijing-funded predecessor 'insignificant in comparison'.

Indian rupees and Chinese yuan crossed the Maldivian shoreline in quantities that echoed the days when money literally washed up on the islands' beaches. In the tenth century, Arab sailors called the coral atolls *Diva-Kauzah* (Cowrie Islands), at a time when cowrie shells (*Monetaria moneta*) were used as currency in ancient China and India. This shell money was later introduced across Africa, attracting the interest of Portuguese,

Dutch, French and English traders who accumulated great quantities for use in the slave trade. For centuries, the Maldives had a near monopoly on the harvesting and exporting of cowries, placing these remote islands at the heart of global commerce. However, the value of a cowrie was related to its rarity, equating to the distance from the source. The closer to the beach where it was picked up, the less it was worth. Few Maldivians got rich from cowries.

In the airport terminal, as bleary-eyed squaddies headed for the British military base at Mount Pleasant, I was given a stern warning about General Galtieri's landmines which, more than six years after the war had ended, still littered the islands. The devices, it seemed to me at the time, contained an explosive symbolism. Regardless of which flag flew over Port Stanley, Argentina's claim upon the land had not been defused.

I checked into the Upland Goose Hotel. Built in 1854 with a white picket fence and substantial conservatory for guests to enjoy their full English breakfast, the establishment could have been transported from the esplanade in one of England's Victorian seaside resorts. It had a homely feel, echoing the architecture and ambience of countless unfussy B&Bs, staring placidly at the dark waters of Stanley Harbour. I was told that, seven years earlier, my room had been one of those commandeered by Argentine troops when they invaded.

My assignment was to tell the story of the economic miracle of the Falklands. After British control had been restored, the UK government unilaterally drew a circle in the sea around the islands, creating what they called a conservation zone. Any vessel wishing to fish in the rich waters now had to buy a licence, a policy that would transform the finances of the territory. I went out with the Fishery Patrol to be filmed holding an Argentine shortfin squid (*Illex argentinus*), a prop that told its own story. The vast bellies of foreign factory ships were crammed with hundreds of thousands of tonnes of fish, I explained to viewers, but virtually none ever landed on the Falklands. The catch sailed away to markets in Spain, Korea, Taiwan and Japan. My squid, representing the most valuable of the species extracted from the exclusion zone, had been sourced from the freezer of a local restaurant.

In 1986, fish licence income was worth more than £29 million, a bonanza for an island community of just over 2,000 people. Within a generation, the Falklands would be listed as one of the richest territories in the world, with a GDP per head twice that of the UK. But my report for the BBC had reflected on the tensions of the new wealth. In the Globe Bar, islanders had complained to me that businesses were being set up with a local front, the squid money ending up in the pockets of foreign investors. 'It's invasion by stealth,' one farmer told me. Just down the street, I had visited an up-market restaurant, the foreign owner boasting how he catered for the South Atlantic champagne set. None of his customers had known the islands before the invasion.

Flying low over the peaty interior in a military helicopter, sheep scattering at the throb of the Chinook, I had jotted down my thoughts on post-war life on the islands. My notes remind me how soldiers on the islands referred to locals as 'Bennies', an offensive reference to a slow-witted and innocent character in a popular TV soap opera. To the islanders, soldiers were known as 'When-I's' because of a tendency to boast of 'when I' was in this or that exotic place. The name-calling was a symptom of a divide between the post-invasion arrivals, who regarded themselves as sophisticated and worldly, and the pre-invasion population, who were protective of their traditional and simple way of life on the island.

The two communities lived side by side, but it became obvious to me that there was not much interaction. They had little in common. For one, the island was a distant corner of the United Kingdom's sovereign territory which needed to be defended. For the other, the island was the centre of their world, where they tilled the soil and buried their dead. It was home.

The distinction is true for many islands around the world. An island's value may be calculated in strategic or economic terms, how it might benefit a distant colonial benefactor or guardian. Such places are often treated like pawns in some grand geopolitical gambit, global power and prestige the prizes at stake. For native islanders, the assessment is generally made the other way around, seeing the island as a place blessed with a unique patina, a local way of life polished by custom and habit. The term 'British Overseas Territory' is revealing in this regard. To the people who live on the numerous islands designated in this way, their territory is not 'overseas'. It is the politicians and officials in London who are across the water.

The people of the Maldives called themselves Dhivehin which, like Îlois, simply means islanders, emphasising the cultural importance of their insular identity. They were proud of their connection to the wider world but also of their independence from it. The Portuguese attempted forcibly to introduce Christianity in the six-teenth century, but Maldivians were fiercely protective of their Islamic traditions, and the end of the bloody but successful campaign to oust the Europeans is still marked by the islands' national day.

First the Dutch and then the British exerted some political control over the Maldives from Ceylon (Sri Lanka), although both colonial powers accepted that it was impractical to impose domestic regulation on one of the most disparate and isolated nations on earth. The ruler of the Maldives styled himself Sultan of Land and Sea and Lord of the Twelve-Thousand Islands, a declaration to the world that, while the landmass amounted to only 115 square miles, his domain laid claim to 500,000 square miles of Indian Ocean. It also acted as a reminder of the strategic value of the Maldives, island garlands draped across vital global supply routes.

During the Second World War, the British established a military base on Gan Island, the southernmost point of the Maldives, its concrete runway later becoming a vital staging post for the Royal Air Force shuttling between the UK and the Far East. The island's value was solely as a stopover to somewhere else – RAF Gan's badge bore the motto 'En Route' beneath a picture of a palm tree on a desert island.

The British had been given access to the atoll as part of an agreement with the Maldivian government that 're-affirmed the position of the Maldives as being under the protection of Her Majesty', with an undertaking that the UK would not interfere in the islands' internal affairs. In the 1950s, 900 villagers were moved out and rehoused on neighbouring Feydhoo Island, the British agreeing to build homes and mosques, as well as providing food for the displaced islanders and a package of economic aid. It was an arrangement, however, that sparked concerns about Maldivian sovereignty and independence. The UK government was accused of forcing the Gan islanders into appalling living conditions with meagre financial support. There were also allegations of 'gunboat diplomacy' after the British High Commissioner to the Maldives repeatedly arrived for talks in the capital Malé aboard Royal Navy warships.

In 1957, the new Maldivian Prime Minister Ibrahim Nasir demanded an end to the building works on Gan Island until the agreement with the British had been reviewed. It was suggested the RAF base might be closed down. All local trade with the UK military was banned, a militia sent to the area to ensure compliance. But Ibrahim had misread the mood on his country's distant southern atolls. The islanders of the Suvadive archipelago had quickly become economically dependent on the UK base. It provided them with work and welfare, improving their quality of life, and the Prime Minister's anti-British rhetoric was regarded as typical of a Malé-based political elite that did not understand the realities of life in the far south.

With anger further fuelled by new taxes on land and boats, rioting broke out and the government militia was forced to retreat into British barracks. On 3 January 1959, after four days of violence, a delegation of island leaders arrived on Gan Island to declare their independence from the Maldives and the creation of a new nation, the United Suvadive Republic (*Ekuveri Suvaadheebu Jumhoorihyaa*). The Maldivian flag was pulled down and the starred tricolour of the USR put in its place. The leader of the uprising and self-proclaimed President of the republic, Abdullah Afeef, placed a letter in the *Times of London* pleading for recognition and support from the British, but the UK government sat upon its hands.

The Maldivian Prime Minister regarded the Suvadive secession as an act of war and instructed his Defence Ministry to assemble a task force to retake the southern islands. Ibrahim himself commanded the *Maldive Star*, a Glasgow-built cargo ship that had seen service in the convoys during the Second World War as the SS *Empire Balham*. Loaded with 700 armed men, mostly volunteers who had answered the call to help, she left Malé harbour on 14 July. After initial skirmishes in rebel-held harbours, the huge show of force was enough to convince the leaders of the coup to surrender, Ibrahim no doubt feeling vindicated as they were arrested and led away.

Support for the USR, however, was more difficult to quash. Three years later, the people of Huvadhu Atoll seceded again, raising the rebel flag once more. Ibrahim had had enough. A second task force of gunboats with government police on board was sent to eliminate the separatist threat permanently, with the Prime Minister overseeing events aboard a yacht named the *Silver Crest*. On 4 February 1962, the fleet anchored off the atoll's capital Havaru Thinadhoo awaiting their orders. There was to be little in the way of mercy. Heavily armed soldiers engaged in the systematic and complete destruction of the town. Hundreds of islanders were killed. Hundreds more were marched into the lagoon until the water reached their necks, from where they watched their homes being razed to the ground, their wells filled with rubble, their crops destroyed and their possessions looted. Rebel leaders and sympathisers were then rounded up and taken back to Malé where they were imprisoned and tortured, many dying from their injuries.

The following year, having finally seen the way the

political wind was blowing, the British gave an ultimatum to the remaining separatists living close to the RAF station on Gan Island. All USR flags were to be unceremoniously removed with only those who pledged allegiance to the Sultanate eligible for employment on the airbase. It was the end for the United Suvadive Republic.

However, one flag was secretly placed in a casket along with a painted coat of arms and other symbols of the now defunct state. The items were taken to a remote spot on Fua Mulaku Island and buried, visiting tourists unaware of the treasures secreted just behind their holiday beach.

Pangaea is observing me. I glance up at my shelf of notebooks and wonder if, perhaps, it has revealed a truth I had not previously seen. Islands do not belong to people; people belong to islands. Territorial sovereignty cannot be decided by the planting of a flag, the signing of a treaty or the deployment of force. It is the consequence of islanders, their endeavours and exploits written across generations, which, like the fossilised marsupial of Seymour, sees them become absorbed into the fabric of the island itself.

CHAPTER 19

ESCAPE

The Psychology of Islands

James MacQueen was a despicable character, but he was to play a positive if inadvertent part in the story of islands. As the manager of a sugar plantation on the island of Grenada in the 1800s, he had a reputation for brutality and cruelty. Satirical placards in his native Glasgow spoke to that reputation.

Wanted Immediately
MANUFACTURING LABOURERS
to go out to the West Indies
3000 Manufacturing Labourers, who will engage to go out
to the West Indies, will receive of course NO wages but all
the kind attention, treatment, comfort, indulgences and
privileges, etc., etc. ... with the addition of a LARGE CART
WHIP, frequently and powerfully to their bare bones ...
For particulars, apply to
James M'Slavery

James was a proud imperialist and vocal supporter of the slave trade at a time when the momentum for the abolition of slavery was becoming unstoppable. He feared that the British Empire's hold upon its island colonies in the Caribbean was in jeopardy, a view encouraged by 'Fédon's rebellion', an ultimately unsuccessful uprising against British rule in Grenada, led by mixed-race French speakers in 1795.

It had been a serious challenge to Britain's colonial authority, and James left Scotland for the West Indies immediately afterwards to help shore up the breach in the empire's defences. He was young and he was angry, helping seek out and execute the rebels who had threatened the traditional order he held so dear. James was appointed manager of the Westerhall Estate in southern Grenada, charged with rebuilding the plantation business and restoring discipline with the crack of a whip. But he feared he was fighting a losing battle, conscious of how revolution in America and France had already reconfigured the foundations of power. He warned that without action, 'ruin swift and inevitable will overtake all our West Indian colonies, and with their fall a deep and perhaps incurable wound must be inflicted on other colonial establishments'.

James returned to Scotland to continue his campaign to maintain the slave trade, editing the ultra-conservative *Glasgow Courier* newspaper and becoming an outspoken defender of the 'West India interest'. But abolition came in 1833 and within a few years, the last of the enslaved people in Grenada had been freed. Like the Caribbean islands themselves, James desperately needed to find new purpose in a post-slavery world.

He sought inspiration from the creativity and invention of the age, in the long-distance steamships being built on the River Clyde which were transforming global trade, but also in a radical pamphlet published in 1837 by a relatively unknown English inventor and teacher. Rowland Hill's paper entitled 'Post Office Reform: Its Importance and Practicability' proposed a Uniform Penny Post for every letter sent, regardless of distance, with an adhesive stamp, the Penny Black, to indicate pre-payment of postage.

James immediately began to think how the approach could be adapted for the colonies. There was a rudimentary international postal system, mail hung on the walls of coffee houses, hooks indicating ships and their destinations, but merchants and traders found it cripplingly unreliable.

Within months of Rowland's plans for a penny post, James had submitted his own proposal for a worldwide network of steamships delivering pre-paid mail to every corner of the British Empire. 'The conveyance of mails and despatches from one place to another is of the utmost possible importance to individuals, and to a country,' he wrote. 'The rapidity and regularity with which such communications can be made, gives to every nation an influence, a command, and advantages such as scarcely anything else can give.' The gum fixing the head of the young Queen Victoria to each letter would be the glue to hold together the monarch's imperial possessions.

It was an immensely complex and ambitious endeavour, but James proved to be just the man for such a logistical challenge, convincing the authorities that they should support his venture. In 1839, the Royal Mail Steam Packet Company (RMSPC) received its charter from the queen, and the following year the Admiralty signed a contract in which James agreed to build a fleet of fourteen steamships for the purpose of carrying Her Majesty's mail, sailing twice a week to the West Indies from Southampton and Falmouth.

Shipyards across Britain were involved in constructing the vessels in record time and to the highest standards. Meanwhile, James personally visited dozens of Caribbean islands, meeting governors and customs officers, arranging coal supplies and docking services, making detailed notes on the facilities at each port.

The first mail bags left for the West Indies aboard the Royal Mail Ship *Thames* on 3 January 1842, but to make the numbers add up, James had loaded the vessel with another valuable cargo – passengers. His calculations suggested that the fares would yield an annual profit of £132,000 for the company, around £15 million in today's prices. To reach that scale would require more than the usual flow of business travellers, however. A man who, just a few years earlier, had been complicit in the misery of transporting thousands of slaves to the West Indies was now seeking to profit from people sailing to the Caribbean for fun.

James was aware that such an idea was at odds with the grim experience of most steam packet travellers at the time. 'The passengers and crews are, with the hatches closed, reduced to the choice, while choked with coal-dust, of being broiled or suffocated,' James wrote. 'Female passengers, in particular, will never willingly undertake, certainly never repeat, a voyage under such circumstances.' It was, James argued, 'a disgrace to England' and proposed instead a first-class service in 'packets of the largest size'.

In 1843, the RMSPC published the first edition of its 'Guide to the West Indies', a brochure looking to woo intrepid travellers aboard the steamships. As well as details of recommended local hotels, transport services and sights worth seeing on each island, the guide used the poetry of Samuel Taylor Coleridge's nephew Henry to evoke the special green and blue of the tropical palette.

> Beautiful islands! where the green
> Which nature wears, was never seen
> 'Neath zone of Europe; where the hue,
> Of sea and heaven, is such a blue
> As England dreams not

The booklet had a refreshing honesty, perhaps fearing the complaints from those passengers who were dissatisfied with the promise of first-class travel with the finest food and wines.

'Everyone reels and totters – there is no rest on deck – to bed with the sick. Discomfort prevails; and many times do some lament they had ever come on board.' But once past the 'heaving billows' of the North Atlantic, the guide's authors suggest a glorious experience. 'To those passengers who for the first time have attempted a long sea voyage, the strangeness of the sailor's life has now become pleasurable. The air is balmy, the pulse quickens, the spirits are alive to all things with a keener relish.'

James believed he could appeal to the more extrovert among Britain's rapidly expanding middle class, adventurers willing to explore beyond their own familiar island shore. Risk was part of the sales pitch.

'It is admitted that at particular seasons the climate in some parts is unhealthy, and from the unfavourable situations of most of the towns, the mortality is great.' But the islands were also portrayed as having a positive impact on mental well-being. 'Nervous diseases have yielded to a residence of more or less length in the islands named,' the guide claimed. The air in Jamaica was said to be 'very light and enlivening, producing great cheerfulness and buoyancy of spirits'.

Henry Coleridge painted the scene on St Lucia. 'The sandy beach stretches like a line of silver round the blue water, and the cane-fields form a broad belt of vivid green in the background. Behind this, the mountains, which extend north and south throughout the island, rise in the most fantastic shapes.'

Although he did not realise it, James MacQueen had stumbled upon an idea that would come to transform the economic and social future of not only the Caribbean but islands around the world – island tourism.

***

My computer pings to alert me to the arrival of an email which has been deemed unthreatening enough to be escorted through my protective firewall. I have placed my security settings at the highest level in an attempt to prevent viruses and villains from breaching my internet defences, a metric for my cyber-islandness, I suppose.

The email is from the Office for National Statistics, a harmless missive updating me on the latest data measuring personal well-being in the UK. The government's number crunchers regularly ask hundreds of thousands of people in all parts of Britain how happy or anxious they feel, how satisfied they are with their lives, how worthwhile they think their existence really is.

I dive into the data, trying to translate the apparently impenetrable matrix of numbers in the spreadsheets into an understanding of the national mood. There is pleasure in this task of sorting and ranking, coaxing the statistics to give up their secret stories. I decide to interrogate every scrap of data going back ten years, thousands of entries representing millions of interviews, to see if where people live has any significant bearing on how people feel. When I examine the results, a remarkable picture emerges. Whichever way you cut it, the cheeriest and most contented places of all are islands.

The Outer Hebrides and the Orkney Islands are the only locations in the country where people give an average happiness score of more than eight out of ten across a decade of data. The same two archipelagos emerge with the highest scores for life being worthwhile. When you rank the hundreds of local authorities according to how their populations rate life satisfaction, Orkney comes top with the Outer Hebrides second and the Shetland Islands in third place. Orkney is also number one in the list of localities where residents feel least anxious.

There are plenty of people who tell me they would be miserable living on a remote Scottish isle, and it may be that those who choose to inhabit such places are a self-selecting cohort of quite particular island lovers. The dark winters, cool temperatures and constant winds are not to everyone's taste, nor the detachment from the hum of metropolitan affairs. Nevertheless, I am convinced the geography of islands does exert a psychological hold upon us, and science offers some backing for this theory.

My own childhood in the west of Scotland was blessed with regular trips to the Isle of Arran in the Firth of Clyde, and I do remember how crossing open water to reach our holiday destination increased my sense of exhilaration. There is something uplifting about a big blue sky. Now living in London,

I often forget how the solid architecture of the streetscape constrains my outlook until I reach the river where the panorama suddenly and thrillingly opens up. It can feel almost overwhelming, a huge gasp of oxygen after the suffocation of urban intensity.

The Ancient Romans understood the theatre of this experience, designing their cities to inspire surprise and wonder, narrow alleys and lanes leading on to wide open piazzas and squares where eyes gaze up as jaws drop down. The drama is contained in the juxtaposition of intimacy and exposure, playing to the introvert and extrovert inside us all. One might say this is the ecstasy of islandness, the glorious tension between private and public, between looking in and looking out.

The effect is even more pronounced when we stare out from the shoreline to where the vastness of the blue ocean meets the vastness of the blue sky. Blue, according to the great German writer Johann Wolfgang von Goethe, contains 'a contradiction between excitement and repose' and there are suggestions this paradox is linked to well-being. Numerous research papers identify significantly lower levels of psychological distress among those who are regularly exposed to the 'blue spaces' of sea and sky. Coastal dwellers who enjoy marine views have lower rates of depression, exhibit reduced stress hormones and report greater happiness than the rest of the population. The impact of green spaces on mental health has been well documented, but science suggests the combination of green and blue is even more beneficial. Some psychologists now prescribe a daily dose of awe for their patients, focusing on something bigger than oneself, explicit in the meeting of the oceans and the heavens. Here, then, is a clue to why islands appear at the top of the UK well-being charts. Their essential physical qualities make us feel better.

On 28 January 1895, the first English cricketers ever to tour the West Indies disembarked from the Royal Mail Ship *Medway* at Bridgetown Harbour in Barbados. The steam packet company had given the team preferential rates, anticipating that the arrival of men in flannels might help improve the image of the Caribbean as a safe and civilised destination. The trip had been the idea of a former Lincolnshire GP and cricket enthusiast, Dr R. B. Anderson, who had been living in Trinidad for the previous twenty-eight years and was keen to dispel some myths about life on the islands. 'The fact is that the ignorance of stay-at-home Englishmen about the West Indies is something appalling,' Dr Anderson told *Cricket* magazine. 'Instead of regarding it as a great health resort, which it undoubtedly is in the winter, people in this country seem to regard it as a special manufactory for ague and malarial fevers.'

The colonial rulers of the Caribbean islands were attempting to market the region as a beneficial escape from the coughs and colds of European and North American winters. Rather than distant lands providing sugar for the tea table, the West Indies were being reimagined as accessible places for those with money to recuperate and relax.

The portrayal still saw the islands as imperial possessions, of course, the black population cast as exotic ornaments in a white fantasy. The clergyman and novelist Charles Kingsley's book *At Last: A Christmas in the West Indies* told his readers what he had learned about the 'negro' people of the islands, observations which epitomised the pervasive racism of the time. 'Their faces shine with fatness; they seem to enjoy, they do enjoy, the mere act of living, like the lizard on the wall. It may be said – it must be said – that, if they be human beings (as they are), they are meant for something more than mere enjoyment of life.'

As the nineteenth century ticked over into the twentieth, attitudes to international travel were shifting. Smoky mail packets were evolving into champagne cruise liners promising luxury and glamour as they ferried high society across the Atlantic. In 1901, the Elder Dempster Line began sailing from Bristol to Jamaica and asked a Scottish doctor who lived on the island, James Johnston, to promote his adopted home. James, who had for some years been lecturing on the health benefits of Jamaica to audiences in Britain and North America, was happy to oblige.

Sailing away to a Southern sea,

Out of the bleak March weather:

Sailing away for a loaf and a play,

Just you and I together.

And it's good-bye to worry,

And it's good-bye to hurry,

And never a care have we,

With the sea below and the sky above,

And nothing to do but dream and love,

Sailing away together.

His book for the shipping line, *Jamaica: The New Riviera*, opened with verses celebrating the still novel idea of going to the tropics for a relaxing winter holiday, lines adapted from a poem by Ella Wheeler Wilcox. 'This book is not offered to the public merely for the purpose of advertising the island,' James explained to his readers, 'but in the confidence that I am thereby fulfilling a duty to tourists, both British and American, by bringing to their notice the existence of a country the possibilities of which, as a health resort, can never be over-estimated.'

His enthusiasm for the island as a holiday destination was quite remarkable, at a time when the West Indies were still seen by many in the UK as backward, diseased and dangerous. 'One of the greatest diversions of the visitor who remains for any time in Jamaica is the sea-bathing, which, at such places as "The Doctor's

Cave", Montego Bay, is said to be the finest in the world,' James wrote. 'The daily sea-bath in mid-winter is at once luxurious and exhilarating. Jamaica is so thoroughly English in social matters that all the English sports are engaged in by the people of leisure, so that one is never at a loss for diversion.'

I am seven years old and skipping along the waterline of Kildonan beach on the Isle of Arran in the Clyde estuary. Little footprints mark the soft sand as I splash through the foam, before gentle waves wash them away again, restoring the intimacy of the private kiss between land and sea upon the shore. Cool water is slapping around my ankles as a tugging breeze throws spray against flushed cheeks. A crab (*Carcinus maenas*) scuttles into the water and I am mesmerised by the creature as it takes evasive action, burying itself in the wet sand of the liminal zone where land meets sea. I am happy.

> I must go down to the seas again, for the call of the running tide
> Is a wild call and a clear call that may not be denied;
> And all I ask is a windy day with the white clouds flying,
> And the flung spray and the blown spume, and the seagulls crying.
> – 'Sea Fever' by John Masefield

Scientists now believe there may be evolutionary logic behind the call of a beach. Having descended from the trees and begun to forage upon the savannah, early man eventually reached the coast to discover a ready supply of seafood, rich in beneficial oils and fats. The crabs, shrimps and other shellfish had high levels of omega-3 and fatty acids, known to increase brain development and improve mood. Healthy snacks could be found simply by lifting a stone in a rockpool.

Science also suggests that the act of walking barefoot on a beach, feeling the soft sand press against the sole and between the toes, is good for us. Research has found that direct physical contact with the surface of the earth has a positive impact on human health. Earthing or grounding promotes physiological changes as the planet's negative electrons are transferred to the body, reducing inflammation and pain as well as improving sleep and helping control some diseases. Kicking off your shoes and paddling in the shallows is more than a simple pleasure.

The sea itself will have been a potential hazard to the hominids on the shoreline, of course, particularly when the syncopated crash of a storm replaced the steady rhythm of the lapping waves. Perhaps that is why medical studies indicate the regular breath of an unthreatening ocean rising and falling on the shingle can help achieve a meditative state that calms and heals and fortifies the brain.

> It is a beauteous evening, calm and free,
> The holy time is quiet as a Nun
> Breathless with adoration; the broad sun

Is sinking down in its tranquillity;

The gentleness of heaven broods o'er the sea:

Listen! the mighty Being is awake,

And doth with his eternal motion make

A sound like thunder – everlastingly.

– 'EVENING ON CALAIS BEACH' BY WILLIAM WORDSWORTH

My parents load my baby sister, two brothers and me into the back of the Morris Minor and tell us we are going exploring. The drive takes us north, through the port of Brodick and past its strange folly of a castle, along Arran's coast road with views across the Firth of Clyde to the mainland. As we enter the little village of Corrie, we pull into a lay-by and my father instructs us to follow him. He leads us onto the pink sandstone beach and issues a challenge to 'look for something interesting'. We skid on the vivid green seaweed (*Ulva intestinalis*) as we clamber over the rocks until, with delight, we come across a curious feature. Steps have been carved into the stone, leading down into a rectangular pit which is filling with seawater as the tide comes in.

'It is the doctor's bath,' my father tells us mysteriously. Four pairs of young eyes look from him to the pit and back again. 'Can you guess why a doctor would build a bath on the beach?' my father enquires. We have no idea. 'Seawater therapy,' he says at last. 'Dr McCredy came to live on Arran about 150 years ago and told some of his patients to bathe in the sea to cure their illnesses.'

'Did it work?' I ask.

'Probably didn't do any harm,' my dad replies, diplomatically.

The idea that the seaside possessed healing qualities had taken hold in around 1750, when a doctor with a practice in Brighton on England's south coast published a Latin dissertation entitled *De Tabe Glandulari* (*Glandular Diseases*). Dr Richard Russell recommended the use of seawater to treat enlarged lymphatic glands and other conditions, a theory which attracted the interest of royalty. The future King George IV stayed in the area to get seawater treatment for gout in 1783. His father, George III, went to Weymouth in 1789 to seek a cure for biliousness, fiddlers playing 'God Save the King' as the royal bathing machine was lowered into the water. It sparked a craze for therapeutic marine bathing that became the inspiration for countless seaside holidays. Suddenly, little coastal villages were transformed into booming resorts as millions sought to escape the choking industry of the town for the fresh air and big skies of the beach.

> A big fat sky and a thousand shrieks
> The tide arrives and the timber creaks
> A world away from the working week
> Où est la vie nautique?
> That's where the sea comes in...
> – 'Nation's Ode to the Coast' by John Cooper Clarke

Evidence that saltwater treatment (thalassotherapy) actually works is hard to find, but the health benefits of being on the coast are well documented. The 2011 UK census asked people to say how healthy they were from poor to excellent, and analysis of the replies from 48 million adults in England shows that the closer people live to the sea, the better they are likely to feel. Coasts are not exclusive to islands, of course, but small island communities have a much more intense relationship with the shoreline and its benefits, both physically and psychologically. The encircling nature of an island coast has an almost mesmerising hold upon those within the boundary line, a thought that transports me back to the Isle of Arran.

Sunlight sparkles on the black waters of the Clyde estuary as my seven-year-old self stares out from Corrie beach, across the enchanted channel separating island from mainland, dividing here from there. The small boy on the shoreline can feel the detachment, conscious of how familiarity lies out of sight beyond the horizon. A combination of trepidation and excitement cause him to tremble as the sea breeze tousles his hair. From this vantage point, it is possible to observe normality, free from its gravitational field. The island provides an escape from the routine, but it is more than that. The seven-year-old on the water's edge feels as if he has broken free of the concentric circles of identity that usually constrain him. He has stepped into a new circle where he can be whoever he wants.

> Ah! Could I tell the wonders of an isle
> That in that fairest lake had placed been,
> I could e'en Dido of her grief beguile;
> Or rob from aged Lear his bitter teen
> – 'IMITATION OF SPENSER' BY JOHN KEATS

Psychologists have been developing the theory of 'vacation identity', an idea that has its origins with Sigmund Freud's assertion that people 'cannot subsist on the scanty satisfaction they can extort from reality'. We need to escape from ourselves sometimes, it is argued, and relocating to an island, even for a short break, provides us with the physical props to do just that. Getting away from normality is the dominant motivation for all tourism, of course, but the act of crossing an island coastline emphasises the sense of entering a new and distinct jurisdiction where the usual rules do not apply.

Physically separated from conventional manners and social norms, island tourists in particular are encouraged to indulge in alternative or extreme behaviours. In the 1970s, resorts on Majorca and Ibiza in Spain's Balearic Islands (*Islas Baleares*) began to cater specifically for holiday hedonism, a culture of excess which flowed along coastal strips of clubs and bars in a flood of cheap beer and pale urine. 'Party islands' sprang up across the Mediterranean and beyond, each selling the promise of oblivion from dreary reality, like Homer's lotus-eaters on their island in the *Odyssey*. Visitors were encouraged to 'let themselves go' in an almost literal sense, to take on a different personality, indulging in risky behaviour that would be unthinkable back home, authorised by the geography of isolation.

> To watch the crisping ripples on the beach,
> And tender curving lines of creamy spray;
> To lend our hearts and spirits wholly

To the influence of mild-minded melancholy

– 'THE LOTOS-EATERS' BY ALFRED, LORD TENNYSON

Islands, as we know, are never entirely separated from the world beyond their shores and consequences inevitably wash up on their beaches. In January 2020, just before the pandemic took hold, Spain's regional government on the Balearic Islands attempted to call time on what was called 'booze tourism', passing new laws banning the bar crawls, happy hours and binge drinking that fuelled the party island fantasy. Other island resorts have also attempted to take back control of their territories from unruly visitors, introducing strict licensing regulations that have sounded the death knell for businesses which sold 'no limits' as their USP.

The character of an island is a product of its islanders but, equally, the character of the islanders is a product of the island. It is a back and forth that sounds to me like the sea advancing and retreating upon the sand.

CHAPTER 20

UNITY

Fun and Freedom

By the early twentieth century, the idea of going to the tropics for fun had seeded itself on the islands of the Caribbean. The year 1901 saw the American oil tycoon and railway magnate Henry Flagler open the Colonial Hotel in Nassau on the Bahamas' principal island of New Providence, a huge wooden structure on the site of the town's Old Fort. To ensure a good supply of guests, he ferried rich tourists over from Florida on the luxury cruise line he had established a few years before. The company's flagship, SS *Miami*, had two tiers of elegant staterooms, panelled in white mahogany and each with electricity, running water and a fan. The orchestra played as the passengers dined, the overture to a winter 'season' on the islands.

A Bahamian postcard from the time showed graceful women in full-length crinolines and straw hats strolling along the beach of Hog Island, just across the water from the Colonial Hotel. In the foreground, one woman was protecting herself from the sun beneath a large umbrella, as black manservants in peaked caps stood guard. In the background, people were frolicking in the sea. There, on the shore of what was later renamed Paradise Island, was the perfect illustration of how the West Indies was being repackaged as a sophisticated destination for the fashionable white elite, the colonial certainties still firmly in place but cheerful modernity skipping along the shoreline.

The First World War slowed the growth of the fledgling Bahamian tourist industry, but just weeks after the guns fell silent, the fortunes of the islands were transformed. The United States ratified an amendment to its Constitution that outlawed the trade in alcoholic drink across America, including a ban on its importation. With Prohibition, the continental power had voted to cast itself as an island of moral rectitude in a sea of vice, attempting to pull up the drawbridge on sin and sinners who might smuggle the demon drink across its borders. But as any student of nissology could have told them, even the smallest island cannot isolate itself from the world, never mind a vast country with a shoreline almost 100,000 miles long. For Bahamians, some living just fifty miles from the US coast, opportunity was knocking.

With its dark history of piracy and smuggling, the Bahamas shrugged off ideas of becoming a genteel, high-class resort and instead established the territory as a base for bootlegging, rumrunning and hedonism. Tens of thousands of barrels and cases of Scotch whisky were imported from Britain, piled high on the dockside at Nassau, then stored in one of thirty-one bonded warehouses on the islands before being secretly trafficked past the continental coastguard. Annual revenue from alcohol rocketed from £81,000 in 1919 to more than £1 million by 1923, a massive windfall that did not go unnoticed in Washington. The Americans complained vigorously to the Foreign Office in London, but officials simply responded with a world-weary sigh, pointing out that the trade was perfectly legal across all the territories of the British Empire.

The Colonial Hotel, destroyed by a devastating fire in 1922, arose from the ashes as the New Colonial Hotel, a beach-side pleasure palace symbolising the islands' ambitions as a refuge from continental puritanism. It boasted 250 'fireproof' rooms, dancing in the gardens and bathing on the beach. The American owner, shipping magnate Frank Munson, was an avowed racist who refused to employ black staff in anything but the most menial roles, advertising his hotel as a place where guests might enjoy 'the society of people of distinction', code for his segregation policies. The Colonial may have been new, but the hotel was still assertively colonial with its whites-only cocktail bars and restaurants. It was an offer which proved intoxicating for many well-heeled Americans from the southern states, who crowded onto the boats and planes leaving dry Miami for the delights of Nassau.

Air services between Florida and the Bahamas had begun in 1919 when Arthur 'Pappy' Chalk started selling tickets for seaplane trips to the Bahamian islands of Bimini from beneath a beach umbrella on Miami's Biscayne Bay. Within a decade, Pan American was running regular services right the way down the West Indies chain to Trinidad, the aviation superstar Charles Lindbergh visiting many of the islands on promotional goodwill visits aboard his *Spirit of St Louis*. As well as hedonists and hustlers, among the pioneering passengers flying out to the islands were keen philatelists collecting the first day covers issued to mark each new airmail connection, commemorative envelopes emphasising the Caribbean's connectedness.

The Bahamas was reinventing itself as an island playground with liberal rules and regulations. Mobsters, swindlers and chancers of all kinds turned up in Nassau, many to make their fortune. The huge profits from smuggling helped pay for improvements to the capital's infrastructure: tarred roads, running water, electricity and telephone wires were installed in the richer white neighbourhoods of Nassau.

The end of American Prohibition in 1933 was a significant blow to the Bahamian economy and the business model of many among Nassau's wealthy residents. The newly installed British governor, Sir Bede Clifford, explained the situation to the islands' Executive Council. 'Well, gentlemen, it amounts to this,' he advised. 'If we can't take the liquor to the Americans, we must bring the Americans to the liquor.' The Bahamas, Sir Bede asserted, now had a straight choice between rapidly expanding its tourism industry or facing bankruptcy.

Attracting wealthy Americans would be easier if the islands maintained an 'all white' policy, the governor suggested to the nodding councillors. Tourist brochures advertised the Bahamas' 'old-world atmosphere', another coded reference to a strict colour bar in the hotels, clubs and on the best beaches. The Development Board overseeing the expansion of tourism was complicit in whitewashing the image of the islands. The Ministry of Tourism's promotional publication, *Nassau Magazine*, rarely showed any black faces on its pages, and postcards of street scenes were reshot with white visitors replacing black locals. 'Profane language', 'the tooting of cabmen', 'music from a dance hall' – the sounds of Bahamian cultural life – were silenced in order to encourage 'a desirable class of tourist to our shores'. When cruise liners arrived in port, extra police were brought in to prevent disabled 'black urchins' spoiling the view. Beggars and other 'annoyances' must be kept over the hill, senior officers were advised, restricted to the shantytowns that lay out of sight behind Nassau's opulent facade.

A century after the last slaves were freed, the black population of the Bahamas was being hidden away like a dirty secret. Beyond the island of manicured gardens and elegant residences in the capital's centre was a sea of extreme poverty where life expectancy was short. A public health specialist sent from London, Sir Wilfred Beveridge, was shocked by the public indifference to the state of Nassau's black districts, where he encountered open privies polluting wells and mountains of festering uncollected garbage. Overcrowding, lack of hygiene, poor nutrition and ignorance were blamed for a litany of preventable diseases. Wilfred's report ended with a warning of an explosive outbreak of typhoid that could be disastrous for the tourist traffic. 'The writing on the wall is not to be disregarded,' he wrote.

The plight of the country's poor black population was largely ignored by the island's white gentry. Voting was restricted to men with property and money, so the descendants of African slaves had almost no voice, even though they made up more than two-thirds of the population. One of the few people to stand up for their rights was the redoubtable editor of the *Nassau Daily Tribune*, a man of mixed French and African ancestry, Étienne Dupuch.

Étienne had inherited the newspaper and a powerful sense of social justice from his father, and the pages of the *Tribune* dripped with a furious outrage that saw Nassau's rich and powerful try to close him down on numerous occasions. In one editorial, Étienne fulminated at how the government 'had almost bankrupted the Treasury to build a hotel, a golf course and provide a steamship line while the people of the Out Islands starve'. He also took aim at the New Colonial Hotel's refusal to employ black staff, prompting a protest that eventually saw the management hire a small number of black drivers to transport white guests over to Hog Island beach.

The expanding tourist industry brought little benefit to the people of Nassau's shanties or the villages on the outer isles. Apart from a few 'native' floorshows and music groups, the jobs in tourism were almost entirely reserved for white candidates, many brought over from the United States. Restrictive covenants banned non-whites from buying prime land and the development of hotels and tourism businesses was managed by a tight clique of Bahamian professionals who worked chiefly with foreign investors in North America and the United Kingdom. The black population could do nothing as their island was being sold off to outsiders.

The New Colonial Hotel changed hands yet again, purchased by an American-born gold mine owner, Harry Oakes, reputedly after he had received poor service there. The immensely wealthy and well-connected tax exile renamed it the British Colonial Hotel, emphasising its white European credentials. He built a new golf course and country club, expanded the airport and invested so heavily in property and land on New Providence that soon he personally owned about a third of the colony's main island.

Harry effectively bought himself a seat on the islands' House of Assembly, crushing his black shopkeeper opponent by getting the local bank secretly to stop his rival's credit and bribing voters with cash and alcohol at the polling station in front of watching police. Bequests of more than $500,000 to St George's Hospital in London resulted in the multi-millionaire being given a baronetcy by King George VI, Harry subsequently able to style himself Sir Harry.

Some developers would hire only white workers and there were near riots when hotel construction projects refused to take on locals. Without employment, many black Bahamians became destitute and desperate, and police reported a surge in begging, theft and arson. Two whites-only theatres in Nassau mysteriously burned down in 1937 and the following year the governor informed the Colonial Office in London that there was a mob of around a thousand 'loafers, criminals and riff-raff' at large in the capital.

The start of the Second World War meant the UK government had other things on its mind than social tensions on a small island colony thousands of miles across the ocean. However, the isolation of the Bahamas suddenly became the answer to a particular problem. The Prime Minister Winston Churchill had learned that Adolf Hitler's invasion plans included installing the Duke of Windsor as a puppet king. The former King Edward VIII, a suspected Nazi-sympathiser, had been exiled to

France since his abdication and intelligence suggested he might be kidnapped and taken to Berlin. To keep him and his wife Wallis Simpson 'out of Hitler's grasp', Winston installed the duke as governor of the Bahamas.

Three-quarters of Nassau's 20,000 population turned out to greet the couple as they arrived aboard the grey cruiser *Lady Somers* on 17 August 1940, the band playing and the streets festooned with Union Jacks. 'Welcome Sweet Papa HRH' read one sign, as the duke and duchess stepped ashore in sweltering heat. In part, the public enthusiasm was a response to global celebrities coming to their small island. But it was also about a hope that having such a high-profile figure appointed to run the Bahamas reflected the concern of the imperial motherland at what was happening in her colony.

It soon became clear, however, that London was much more interested in the islands as a bargaining chip with Washington than the plight of those trying to survive in Bahamian slums. As part of a deal for fifty US destroyers, the British agreed that the Americans could construct two military bases on New Providence Island, one of them at Oakes Field, the airport owned by Sir Harry.

The building project did bring much-needed employment, of course, but local labourers were dismayed to discover that imported American workers were earning double what they were paid for the same tasks. Dismay turned to fury when they learned that the construction company had wanted to pay black workers a higher rate but had been prevented by their own Bahamian government. The conclusion seemed unavoidable. The white government was deliberately keeping the black population poor and uneducated to prevent it gaining any foothold on power.

On 1 June 1942, thousands of black workers and their families marched to Nassau's Bay Street, the main artery of Bahamian commercial and political power, and began smashing windows and looting stores. A parked Coca-Cola truck, a symbol of American commercial infiltration, became a source of missiles to hurl at police and the soldiers despatched from the local garrison. Rioters sang 'We'll Never Let the Old Flag Fail' as one man, Napoleon McPhee, burned the Union Jack. 'I willing to fight under the flag, I willing even to die under the flag, but I ain't gwine starve under the flag,' Napoleon said.

The Riot Act was read and a curfew imposed, but unrest continued through the night and the next day. Fire engines and ambulances were overturned and the iron bell that sat on top of the Southern Police Station was pulled down and placed in the belfry of St Agnes church, 'to summon saints to worship instead of officers to arrest the poor', it was later explained.

When order was eventually restored, four protesters lay dead, as the Duke of Windsor blamed 'communists' and 'men of Central European Jewish descent' for being behind the trouble. The unrest had knocked the self-confidence of Nassau's ruling class, who knew how uprisings and strikes had been sweeping across other islands in the Caribbean. 'I never thought that our people could be agitated to the point of rioting, because they have always enjoyed the enviable reputation of being patient, docile and law-abiding,' Étienne Dupuch lamented.

Étienne didn't yet know it, but Bahamian unease was about to deepen with the biggest scoop in the history of the *Tribune*.

On the night of 7 July 1943, a violent tropical storm raged in the skies over Nassau, thunder booming as lightning illuminated sheets of torrential rain. At some time after midnight, an intruder entered the bedroom of Sir Harry Oakes at his mansion on Cable Beach and struck him four times behind the left ear with a miner's hand-pick, killing him as he lay in his bed. The body was then doused in insecticide and set alight, after which the murderer sprinkled Harry's corpse with white feathers from a pillow.

When the telephone at the Oakes residence rang around breakfast time, it was Étienne checking his scheduled meeting with Harry that morning was still on. The call was answered by Sir Harold Christie, a business associate and friend of Harry's who had stayed overnight. He told the newspaper man that he had just found the bludgeoned body of his host, feathers still being gently blown around the bedroom by the ceiling fan. Étienne was already designing his front page.

The murder would become one of the great unsolved

crimes of the twentieth century, spawning numerous books and four feature films. Étienne, aware that the islands' leaders would want to downplay the incident, ensured it was on the front page of newspapers around the world, much to the fury of the Duke of Windsor. 'It must be remembered that Dupuch is more than half negro,' the duke complained. 'Due to the peculiar mentality of this race, they seem unable to rise to prominence without losing their equilibrium.' The duchess commented that there was 'never a dull moment in the Bahamas'.

Étienne's campaigning fire had been stoked and, along with his lawyer brother Eugene, he pledged to fight for the rights of the black population of the islands. His crusade would culminate in an extraordinary night of drama in the House of Assembly on 23 January 1956. Étienne had tabled a resolution deploring racial discrimination in the colony's hotels, theatres and everywhere else that imposed a colour bar, describing segregation as 'not in the public interest'.

The proposal was deeply unpopular with the island's powerful business and political leaders, who believed it would damage the tourism industry and assumed the assemblymen would consign it to the long grass. But with the public galleries packed, Étienne rose to his feet to make a famous speech. 'Who of you tonight is prepared publicly to declare that a whole group of people should continue to live a life of daily and constant humiliation because it might mean the loss of some material gain?'

he asked. 'The day is past in the world when classes and races can be divided by some cruel invisible line.'

When it was declared the resolution would be sent to what everyone knew was 'the graveyard committee', Étienne refused to sit down, despite a threat from the Speaker to have him arrested. 'You can call the whole police force, you can call the whole British Army, I will go to jail tonight, but I refuse to sit down,' Étienne proclaimed.

'Don't touch him!' the popular Bahamian musician Freddie Munnings shouted from the gallery.

The meeting was hurriedly adjourned, and Étienne left the building surrounded by a large crowd of supporters. Within days, the islands' hotel managers announced they would end racial discrimination in their establishments. Sir Harry's widow, Lady Eunice, took out a full-page advertisement in the *Nassau Daily Tribune* to confirm that the British Colonial Hotel was now open to all.

I am back. More than half a century since I was last here as a boy, I have returned to the Isle of Arran, Pangaea asleep in my jacket pocket. On Kildonan beach, I remove my shoes and socks and paddle in the cold, clear water. For old times' sake. The little cottage where we stayed each summer is just as I remember it. There is the dormer window behind which I used to stand on tiptoe to see the sea, to wonder at the lighthouse on little Pladda Island, and to stare at the enigmatic profile of Ailsa Craig on the horizon. Seeing the line of the hills and the sweep of the path that arcs down to the shoreline is like being reunited with old friends. My father's ashes were scattered nearby, the spot marked with a cairn of pebbles. Ceremoniously, I add another stone to the pile. This is a landscape haunted by memory and meaning.

I kick at the surf as if to break a spell. Such places have a power over us, and island landscapes are possessed with a particularly strong form of magic. Humankind has always been drawn to them and, in recent times, islands have wooed travellers with mystical promises of healing and restoration. My palm cradles Pangaea in my pocket as I walk up the hill to the bus stop and a journey along the coast road.

Understanding the lure of islands means unpicking what it is about the physical geography of a place that affects us psychologically. Landscapes which we find beautiful or affecting generally turn out to be battlefields of contradiction, where unity squares up to complexity and legibility confronts mystery. On islands, the effect of these paradoxes is often amplified by compression, the ABC of islandness at work.

Unity is the binding force that successfully holds the different aspects of an environment together. Watching Arran's shoreline roll past the bus window, I realise that, for islands, unity comes from the encircling coast, relentlessly emphasising the isolation and separateness of the place. The structural definition of the territory is reassuring, a distinct boundary that gives the space within an appealing integrity and identity.

Complexity, however, is the flip side of the same coin. There can be a melancholy splendour in a featureless desert landscape, or the stunning homogeneity of vast prairies, but our eyes are always searching for interest, and in time, if we find none, disappointment or despair may overwhelm us.

Arran markets itself as 'Scotland in miniature'. Like the mainland, it is riven by a geological fault line, with highlands to the north and lowlands to the south, dark forests and lush pasture, moors and glens – all within a coastline of just ninety kilometres. 'The Isle of Arran is a place where you can find a little bit of everything,' the brochures boast. As William Cowper's famous poem puts it: 'Variety's the very spice of life, that gives it all its flavour.' The Italian philosopher Umberto Eco goes further, reflecting on how 'the beauty of the universe consists not only of unity in variety but also of variety in unity'. Unity without complexity is tedious. Complexity without unity is chaos.

Another paradox in what psychologists call the 'ideal environment' is the requirement for both legibility and mystery. Legibility relates to our need to understand the topography, to comprehend its scale and successfully navigate its terrain. Strange landscapes make us uncomfortable, and the urge to conquer mountains or traverse oceans surely comes from an evolutionary pressure to control the territory.

Size matters. If a place is too small, we fear there is nowhere to hide, too big and we fear getting lost. Islands, with defined boundaries and dimensions, provide clarity in our search for a happy balance between personal autonomy and social collaboration, between the private and the public. The place where 'I' meets 'us'.

On most islands, the shoreline is a constant presence, always there to reorient us should we lose our way. The dry-stone walls and neat hedgerows which mark the boundaries of Arran's fields make the landscape more intelligible and pleasing. It is, perhaps, the primeval fear of wilderness that means we appreciate elements of structure and order in a landscape. We can relax.

The Yang to this Yin is mystery. A straight road heading to the horizon may be comprehensible, but a winding lane where one cannot see what might be around the next bend is much more exhilarating. We need hills to climb and seas to cross. We need obstacles to overcome. We need questions to answer. That is what gives us purpose.

As the bus continues along its coastal route, Goatfell Mountain brooding to my left and the sea coyly offering glimpses of itself through drifts of yellow gorse on my right, I take Pangaea from my jacket. She is dreaming on her low couch, as serene as one of the stone-grey seals (*Phoca vitulina*) sunbathing on an Arran boulder. Contentment is to be found in the place where opposing forces meet, the shoreline where the solidity of the known gives us the platform to take on the challenges of the unknown.

CHAPTER 21

VISITORS

Fantasy and Truth

Dozing beneath palm trees on the powder white sands of Jolly Beach in Antigua, tourists must have dreamed they had discovered their true island paradise. It was the mid-1990s and elegant yachts drifted into nearby Jolly Harbour, bringing movie stars and millionaires to an unspoiled tropical hideaway, a setting so picture-perfect it could have been straight out of a Hollywood production of *Robinson Crusoe*. The pleasure-seekers applying their sun cream and sipping their piña coladas were not to know that the scene did indeed owe more to the artifice of the film set than the ancient geology of the Lesser Antilles.

The beach was not authentic, nor were the palms, nor the welcoming embrace of the little harbour. This was a simulacrum of a tropical island paradise, designed by Europeans to replicate what they thought a Caribbean holiday resort ought to look like.

Less than a decade before, Jolly Harbour had been a mangrove swamp, home to West Indian whistling ducks (*Dendrocygna arborea*), white-cheeked pintails (*Anas bahamensis*) and Caribbean ospreys (*Pandion haliaetus ridgwayi*). It had been a playground for generations of local children who used to enjoy teasing the birds with their catapults and collecting mangrove oysters (*Crassostrea rhizophorae*) and marsh crabs (*Ucides cordatus*) from the brackish waters.

Mangrove swamps have been described as the rainforest by the sea, an endangered ecosystem teeming with life in the nutrient-rich intertidal zone where the fresh waters of the land meet the salt of the ocean. The wetlands of Jolly Bay in western Antigua were a rare and precious island habitat.

But in 1989, the bulldozers and earthmovers arrived. The scream of chainsaws startled the flocks of black-necked stilts (*Himantopus mexicanus*) as the demolition gangs began their work, slicing through the tangle of red mangroves (*Rhizophora mangle*), dredging the seagrass beds and smashing the coral reefs.

Local villagers were horrified at what was happening to their shoreline. The previous summer, hundreds had presented a petition to Antigua's government protesting at the wanton distribution of their coastal margin to 'foreign adventurers'. The Prime Minister, Vere Bird, dismissed the complaints, arguing the swamp was useless

land which should be redeveloped to bring economic
benefit to the islands.

Since full independence in 1981, control of Antigua
and Barbuda had been dominated by the Bird family, a
political dynasty which successfully guided the islands to
freedom from colonialism but which became mired in allegations of corruption and criminality. Vere, along with
his sons – Vere Junior, Lester and Ivor – had encouraged
the almost untrammelled expansion of international
tourism to the islands, accused of syphoning off personal
fortunes from the tide of foreign cash that washed up on
Antiguan beaches.

The Jolly Harbour project was the brainchild of a
Swiss multi-millionaire, Dr Alfred Erhart, a travel tycoon
who owned the 500-room Jolly Beach Hotel. He was
looking to take advantage of the Antiguan development
boom by building a resort around his hotel that would attract deep-pocketed visitors from Europe and the United
States. Alfred explained to the Prime Minister how he intended to replace the 'mosquito-infested swamp' with a
Florida-style marina, part of a gated community of villas
and condominiums offering exclusive access to foreign
paying guests. Vere was delighted to support the project.

Behind the barriers erected around the construction
site, an authentic and vibrant piece of Antigua's shoreline
was destroyed for ever. In its place, an artificial geography was manufactured. Thousands of tons of sand were

brought in to make a beach, the white dunes of Barbuda, forty miles across the sea, plundered for the raw materials. Horrified Barbudans protested to the government as the excavation contaminated the island's freshwater aquifers, leaving some supplies unfit to drink. Their anguish fell on deaf ears, not least because the man who controlled the sand-mining company, SandCo, was the tourism minister and son of the Prime Minister, Lester Bird.

Another marina and luxury apartment project a few miles north of Jolly Harbour, financed with Italian money, proved equally controversial. Construction at Dickenson Bay required an ancient mangrove basin and salt pond to be dredged and filled. There was a local outcry, with posters and handbills demanding politicians 'Save McKinnon's Swamp', but to no avail.

One morning in June 1989, tens of thousands of dead fish were found floating close to the bay. Experts concluded they had been killed by the destruction of the coast's natural circulatory system and raw sewage being pumped into the salt pond from nearby hotels.

The integrity of Antigua's shoreline was being sacrificed in an almost frantic effort to attract international investment and foreign tourists to the islands. To finance the construction boom, the government borrowed hundreds of millions of dollars, becoming heavily indebted to US and European bankers. The luxury resorts springing up along the coastline were generally foreign owned and managed, much of their profits disappearing overseas.

By the mid-1990s, around two-thirds of Antigua and Barbuda's economy was based on tourism, with half the islands' workers servicing the needs of foreign visitors, overwhelmingly in junior roles.

Newly won independence still appeared reliant on an old colonial hierarchy. Wealthy white plantation owners had been replaced by wealthy white holidaymakers, 'the newly wed and nearly dead' as hoteliers described the clientele coming to play on the pleasure periphery.

Meanwhile, environmentalists warned that the beaches which attracted those tourists were literally eroding into the sea, coral reefs were dying, wildlife was declining, the authentic Antigua was being lost. 'Antigua: the Caribbean as you've always imagined it' became the department of tourism's international marketing slogan. The islands were selling themselves as a white fantasy.

It was a similar story in island states scattered across the Caribbean Sea, former colonies struggling to define themselves on their own terms. The Trinidadian-born writer V. S. Naipaul described the territories as 'manufactured societies, labour camps, creations of empire' in which dependence had become a habit. Centuries of British divide-and-rule policies in the region had helped create highly insular communities, and London's hopes of building a West Indian Federation in 1962 were to be dashed upon the rocks of inter-island rivalry and suspicion. Although the histories and challenges of the islands bore great similarities, the islanders of the

Caribbean tended to focus on their differences, the aspects of each island experience that set them apart from their neighbours. Mighty Sparrow, a calypsonian from Trinidad, sang of the failed attempt to create a West Indian identity.

> Federation boil down to simply this:
> It's dog eat dog, and survival of the fittest.

The St Lucian poet Derek Walcott also reflected on the scattering of emergent island nations in the Caribbean Sea.

> Open the map. More islands there, man,
> than peas on a tin plate, all different size.

The prize of island independence gasped for breath beneath the pounding waves of global economics. Caribbean sugar, banana, coffee and cocoa farmers were exposed to international markets and struggled to keep their heads above water. The United Nations and World Bank implored the island governments to diversify from the single-crop model of the plantations by expanding their tourism industries. But, as in Antigua, the efforts to woo American and European travellers came at a price. Every charter flight that landed, every cruise liner that docked, every foreign-owned resort that opened, risked bleaching the colour from the authentic character of the islands, like postcards left out in the tropical sun.

Meanwhile, the sedimentary residues of slavery had become fossilised into the strata of island societies, a shaky geology that made the challenges of nationhood even more precarious, rumbling resentment and structural inequality destabilising efforts to build unified identities.

In Jamaica, leaders feared their island's personality was being subsumed by simplistic caricatures, portrayed as either sun-kissed paradise or crime-ridden hell. 'The national brand should not be reduced to myths,' ministers agreed, resolving to establish a vision of Jamaica that was more nuanced, multidimensional and accurate. In 2004, they invited a leading British branding consultant, Simon Anholt, to root around in the island's past and present and see what he could find. His report, written in the language of marketing, landed on their desks two years later.

'Jamaica', Simon told them, 'is sitting on a treasure-house of natural brand equity', with 'amazing quantities of intellectual capital in music, art, religion and business'. The problem, he explained, was that the proceeds from this 'vigorous and abundant' resource had been haemorrhaging out of the island for centuries. 'It is absolutely necessary for Jamaica to be able to effectively and comprehensively manage its nation brand to ensure that the brand image is true, fair, up to date and beneficial to the nation,' Simon's report concluded.

Officials looked for evidence that their island's personality was being exploited by outsiders and were alarmed to discover that the word 'Jamaica' was included in hundreds of registered trademarks in Britain and North America, the vast majority of which had no connection or affiliation with their country whatsoever.

The green and gold of the island's flag had been appropriated by global corporations, stitched into trainers, printed on T-shirts and stuck on bottles, marketing shorthand for Caribbean cool. Not only was their culture being commoditised but the profits from this identity theft were ending up far across the ocean.

The fierce competition to attract tourists had seen Caribbean governments focus their marketing efforts on almost identical promises of sea, sun, sand and sex, the '4S model' as the industry described it. As a result, tourists tended not to differentiate between the island resorts except by affordability and accessibility. But around the turn of the millennium, the thinking changed. In part, it was a business decision to seek some unique selling point, but it was also culturally important for each island nation to proclaim its individual story.

In the 1990s, a dozen of the region's destinations used the word 'Caribbean' in their tourist slogans. Saint Croix in the US Virgin Islands advertised itself as 'America's Caribbean Paradise', Martinique was 'The French Caribbean Haven' and Aruba was 'Paradise Found in the Dutch Caribbean'.

Within two decades, both regional and colonial adjectives had disappeared from the slogans. Instead, marketing emphasised island distinctiveness. Saint Croix had 'A Vibe Like No Other'. 'There's Only One Martinique' was the new tagline of the French island, a destination with 'a cachet all its own'. Aruba was 'One Happy Island', with a geography that 'sets us apart' from anywhere else in the Caribbean.

The beaches were still the big draw, of course, but the marketing strategies sought to highlight often subtle points of difference in island tradition, notably the music. Trinidad and Tobago was the home of calypso. In Barbados, it was spouge. Martinique lay claim to chouval bwa. Curaçao had its tumba while Guadeloupe danced to gwo ka. Each island proclaimed its own musical heritage, a unique blend of influences, with roots in the plantations and echoes from ancestors across the ocean. In Jamaica, of course, it was reggae, and the global superstar that was Robert Nesta Marley.

> We know where we're going
> We know where we're from
> We're leaving Babylon
> We're going to our Father's land
> – 'Exodus' by Bob Marley

To many descendants of slaves in the Caribbean, island identities had become intertwined with their continental African heritage, a black consciousness that had found a voice in Pan-Africanism and a creed in Rastafarianism. Encircling coastlines represented both the embrace of belonging and a shackle of enslavement. The poetry of calypso and reggae celebrated island life but also sang of freedom from suppression and a journey to the promised land.

Bob Marley's extraordinary international success encouraged greater self-confidence across the Caribbean, a recognition that from the pain of displacement and suppression, from the struggle of islanders across centuries, something special had emerged. It was not European, North American, nor African. It was a unique product of the islands with global resonance.

Bob's image appeared on numerous Jamaican postage stamps, including a $100 edition to mark what would have been his fiftieth birthday. His face was also reproduced on the stamps of Grenada, Montserrat, Tanzania, Togo, Niger, Chad, Mozambique, Burundi, Djibouti, Burkina Faso, Guinea, Myanmar and Mongolia. Arguably, apart from British monarchs, no individual's portrait had been reproduced upon the philately of so many states around the world.

In January 2020, just before the pandemic struck, Jamaica's Tourist Board announced a new brand positioning for the island. It was revealed that the tagline would be 'Jamaica: Heartbeat of the World', representing 'the beat that makes the world move'.

'On the map, Jamaica may seem like a small dot in the Caribbean Sea,' reporters were told. 'But Jamaica's influence on the world culture is the size of a continent.'

—————————————— ✿ ——————————————

A circuitous journey across land and sea has brought me to the village of Port Charlotte on Islay, the southernmost island of the Hebrides. Pangaea is with me as I sit on the jetty wall and watch the dunlins dabbling in the margins of Loch Indaal. The water is calm and reflective, matching my mood. White-washed cottages on the green banks of the bay lead my eye to the little lighthouse guarding the harbour, as a small boy builds a sandcastle on the beach.

It is an untroubled scene. People come to the islands for the simple life, to go off-grid and turn off the noise. But it is more than escaping the tyranny of emails and social media. They are searching for something reassuringly straightforward and comprehensible. There is good evidence that people who unclutter their lives tend to be happier and more contented, and perhaps that is the most compelling reason for Scottish islands sitting at the top of the UK's well-being charts.

Marvelling at the beauty of the view, the cloudless sky mirrored in the waters of the sea loch, the gentle contours of the opposite shoreline painted soft shades of violet and blue along the horizon, I find myself returning to that balance of unity and complexity which is said to be required for the ideal environment. Human beings are fascinated by the notion that, at the heart of existence, is something or someone that answers every question. Both religion and science propose a unifying theory of everything, a proof or creed that links and makes sense of the wonderful incomprehensibility of the universe.

Philosophers and priests have been drawn to islands as retreats for thousands of years, places to think about the meaning of life, sheltered from the distractions of the day to day. It is a tradition that can be traced back to ancient China and ancient Greece. Irish monks arrived in Islay in the sixth century AD, around the time St Columba was planting the seeds of Christian communities on other Scottish isles including Iona and Inchcolm. Many small islands have tales of mystics and hermits who sought solitude and separation in a cave or grotto, cutting themselves off from earthly demands in the hope of connecting with something bigger. I glance at Pangaea, who appears to be smiling.

This version of events does not see islands as peripheral, but rather as a key in humanity's endless search for truth. Isolation is an essential part of achieving that spiritual link with something beyond ourselves. Gautama Buddha, Jesus Christ and Muhammad ibn Abdullah each found personal enlightenment in places separated from the immediate bustle and hurry of ordinary life. They drew a boundary around themselves, creating an island of contemplation from which they pursued their path to heavenly paradise.

I pop Pangaea into my pocket and walk up the hill to an old chapel built from the charity of islanders who hoped it would lead them to their own Eden. The Kilchoman Free Church, surrounded by the graves of the faithful, is no longer open for worship. Today, surrounded by the island's dead, the building is the Museum of Islay Life, documenting the experiences of the island over the generations, filled with objects, books and photographs donated by local people. Cannon barrels and spinning wheels, milk churns and ships' bells, Mesolithic flints and Victorian mangles, ferry tickets and turf spades – an eclectic jumble of exhibits, each carefully labelled and catalogued, combining to tell the story of an island in all its astonishing variety.

One object catches my eye, an illegal whisky still that was first hidden from the authorities around two centuries ago. Once the islanders' secret, local whisky is now Islay's global claim to fame. With nine active distilleries and more planned, Islay is 'whisky island', its world-famous selection of single malts sufficient to see it designated one of the five official Scotch regions. Tens of thousands of visitors come to do whisky tours and tastings each year, an invasion that far outnumbers the local population.

The single malts of Islay are said to bring together the unique characteristics of island life: peat smoke and sea spray, barley and bonfires. Aficionados travel thousands of kilometres to taste the pure spirit of the island, literally and metaphorically. I am not expert enough to identify all the subtle aromas in a dram of Laphroaig or Bunnahabhain, but I understand the attraction. There are countless layers of flavour distilled into one simple shot. A single malt has infinite layers.

Visitors tick off the distilleries as they travel around the island, collecting the different whiskies until they can say they have the complete set. Watching the tour buses, I find myself reflecting on how we seek to distil the unfathomable into the comprehensible. The human brain is simply not large enough to understand the true meaning of infinity, but we cannot deny its existence. So, we come up with ways to disguise our limitations and soothe the agony of the impossible. We seek understanding by applying structure to chaos, categorising and systemising the world around us. The Linnaean taxonomy which peppers this book reflects the need to balance complexity and unity, drawing boundaries around things, creating distinct and comprehensible identities. That same desire explains the pleasure I get from a set of postage stamps or an alphabetical index. It helps us feel we are getting closer to the truth.

The lure of islands is, in part, that they appear easier to understand, their convolution contained within defined and distinct boundaries. Many newcomers to rural islands see themselves as refugees from big cities, where the chaos of the human experience is written in neon lights. They want to escape the forces of speed and mass that can define urban life. This, too, is an illusion, because we are not able to pull up the drawbridge on the infinite; we cannot run away from the truth. Everything and everywhere is much more complicated than we can ever comprehend.

In the early 1930s, Aimé Césaire, a poet from the island of Martinique, coined the word '*négritude*' to describe a shared identity of black people around the world. It was a term that had a resonance for many people of African descent who lived on former slave islands in the Caribbean and the Indian Ocean, a sense of belonging and connection to continental forebears.

But within three decades, another Martinican writer, Édouard Glissant, rejected *négritude* as too monolithic and simplistic. It did not take account of the intricacies of island identities, that black islanders were no longer African descendants but had become 'new beings in a different space'. Édouard, with others, suggested the term *'créolité'* – a multifaceted identity that was 'an annihilation of false universality, of monolinguism, and of purity'. Unity was found in complexity, and that was a source of pride. 'Neither Europeans, nor Africans, nor Asians, we proclaim ourselves Créoles,' Édouard said.

I am sitting with Pangaea watching the tide go out on Islay as the afternoon ferry heads for the mainland loaded with day trippers. It almost feels as if the island is sighing with relief as the aliens depart. Not that they aren't welcome, but islanders are protective of their precious separation and the distinctiveness they believe comes with it. In whisky terms, they regard themselves as a rare single malt rather than one of the ubiquitous branded blends that can be found everywhere. The fact that virtually all of Islay's distilleries are owned by multinational companies, and that most of the profits from the malts leave the island, causes some consternation among locals. There is concern, too, that the uniqueness of Islay is threatened by the forces of globalisation, a creeping homogeneity that blurs and smudges the island's integrity, its culture and tradition commoditised and packaged for the international market.

These are the inevitable conflicts of the inter-tidal zone,

the fuzzy juncture where 'us' meets 'them'. The tide will always come back in again with fresh influences and impacts, the seedpods of infinity washing up with every wave that breaks upon the shore. The island still has a quality that is unique, of course. There is nowhere quite like it. But unity is not the same as purity. Perhaps we should all proclaim ourselves Créoles.

CHAPTER 22

ISLANDS

Completing the Circle

On 8 July 1831, Francesco Trifiletti was sailing his brig-schooner, *Il Gustavo*, from Malta to Sicily. He had left Valletta three days earlier heading for Palermo, a journey he had made countless times before. Visibility was good with a manageable wind and Francesco was on deck when he noticed something strange about ten miles from his vessel. He stared at the phenomenon almost disbelieving what he was seeing – a miracle (*un miracolo*), as the Sicilian captain would later describe it. The waters of the Mediterranean had broken. An island was about to be born.

On his arrival in Sicily, Francesco told the authorities how he had seen a column of water rising 100 palms (eighty feet) into the air and had changed his course to get a closer look. As he approached, there was a thunderous roar and the column rose again, a tower of black water that remained for ten minutes before gurgling and plunging back down in a cloud of dense smoke. The phenomenon was repeated every twenty to thirty minutes, eruptions that killed hundreds of fish, their bodies scattered on the agitated waters attracting immense shoals of porpoises and multitudes of seagulls.

The location was marked on the charts, a spot halfway between the furthest Mediterranean ports of Tangier and Beirut and some thirty miles south-west of the Sicilian coast. Something decidedly odd was happening in the very middle of the middle sea.

Two days after Francesco had seen the water plume, Sicilian naval commander Giovanni Corrao was in the same area and reported seeing a huge column of water rising perpendicularly from the sea, surrounded by sulphurous smoke, and accompanied by violent thunder that 'added to the grandeur and novelty of the scene'. News then came in that the skipper of a small sailing boat had watched 'a large rugged island, coming up and falling with force back into the sea', and when Giovanni returned to the location a few days later, he saw, through the smoke, the unmistakeable outline of a tract of land, virgin territory emerging from the deep.

The explosive events in the Mediterranean sent shockwaves around Europe, triggering a scramble for control of this new island in a vitally important strategic position. When the British naval commanders in Malta heard what was happening, Vice-Admiral Sir Henry Hotham immediately despatched a cutter to the area, and then a much larger brig was sent as a show of force. Sir Henry didn't know it, but another British ship returning to Valletta from Marseille was already on the scene.

Commander Charles Swinburne, aboard the sloop *Rapid*, had changed course after seeing a column of ash illuminated by brilliant flashes and the light of the moon. 'Several successive eruptions of lurid fire rose up amidst the smoke,' he reported. 'At 5 a.m., when the smoke had for a moment cleared away at the base, I saw a small hillock, of dark colour, a few feet above the sea.' Charles climbed into a smaller boat and headed towards the island, the water around him discoloured with dark objects. 'The crater (for it was now evident that such was its form) seemed to be composed of fine cinders and mud of a dark brown colour; within it was to be seen, in the intervals between the eruptions, a mixture of muddy water, steam and cinders, dashing up and down.'

He estimated the crater to be seventy or eighty yards across and between six and twenty feet high. 'Suddenly the whole aperture was filled with an enormous mass of hot cinders and dust, rushing upwards to the height of some hundred feet with a loud roaring noise,' he told Sir Henry in his report. 'Renewed eruptions of hot cinders and dust were quickly succeeding each other, while forked lightning, accompanied by rattling thunder, darted about within the column, now darkened with dust and greatly increased in volume, and distorted by sudden gusts of whirlwinds.'

Before the dust had settled, the Royal Navy ordered Captain Humphrey Senhouse to claim the new volcanic island for the British Empire. On 3 August, he landed on the smoking crater, hoisted the British ensign, gave three cheers and named the territory Graham Island after Sir James Graham, First Lord of the Admiralty.

However, as described in the official report to King William IV, 'the patriotism of the Sicilians was highly excited by this achievement within the sight of their shores', and those sons of Etna 'retook what they conceived to have been Nature's gift to their sovereign and planted the flag of the Kingdom of the Two Sicilies upon the island'. The volcanic cone that had risen above the waves was renamed *Isola di Ferdinando II* (Ferdinandea Island), in honour of the Sicilian monarch.

A month later and a third European power was claiming the island. The Parisian geologist Constant Prévost led a team of scientists to the crater, climbed to the highest point, now estimated at 250 feet above sea level, raised the French tricolour and rechristened the island *Île Julia*, after the month of its birth. The Spanish, not wishing to be outdone, also suggested the island belonged to them.

As its fame spread, other names for the sulphurous smoking horseshoe of basalt appeared: *Corrao*, after Captain Giovanni; *Sciacca*, after the nearest town; *Hotham*, after the vice-admiral; *Nerita*, after a nearby sandbank.

Whatever it was called, the island was destined to

suffer an early death. Just a hundred days after first appearing from the foaming waters of the Mediterranean, the volcano gave a last gasp and slipped below the surface once more, taking the various territorial claims down with it.

'As the isle was visible for only about three months,' the British geologist Charles Lyell noted wryly, 'this is an instance of a wanton multiplication of synonyms which has scarcely ever been outdone even in the annals of zoology and botany.'

The island's short life was important scientifically, however, used to shore up the controversial Doctrine of Uniformity – the assertion that the laws of physics have always applied and will always apply everywhere in the universe. In geology, uniformitarians believed that the shape and character of land and sea was the result of an even-paced gradual evolution that meant the present was both the key to the past and, by extension, the future. Gradualism was a challenge to catastrophism, the theory that the features of the earth's crust could be explained by sudden violent events, like the great flood as described in the Bible and other creation myths. Today, the foundations of geology involve elements of both theories.

Gradualists argued that Ferdinandea Island (the name that finally stuck) had appeared many times before and was probably destined to emerge again. It was reported that the volcano had broken the surface during the First

Punic War in the third century BC, subsequently appearing and disappearing four or five times.

In 2002, volcanologists reported renewed seismic activity around the site, with speculation that Ferdinandea might be about to affect another entrance. To forestall any new claims of sovereignty, Italian divers swam down to the summit of the submerged volcano and planted the national tricolour. Sicilian divers did the same, fixing their island flag (*Bannera dâ Sicilia*) to the rock, along with a plaque that read: 'This piece of land, once Ferdinandea, belonged and shall always belong to the Sicilian people.'

They were statements of territorial ambition, of course, but there were echoes in the divers' actions of something mythical and magical and ancient. It was as though they were laying claim to Atlantis, seeking to restore the island utopia that had been taken from humanity when it sank beneath the waves in that terrible night of fire and quake. It was about paradise lost and the dream of paradise regained, where flame and rock and water cavort, right in the middle of the middle sea.

Beneath a grey and heavy sky, white spray smashes onto the black sand of Reynisfjara beach. I am buttoned up against the cold wind, sheltering beneath towering columns of geometric basalt, Pangaea nestled in my pocket. It is a scene of funereal monochrome. 'Ashes to ashes, dust to dust,' I recite to myself, boots sinking into the sooty remains of a long-forgotten battle between fire and water.

On the southern coast of Iceland, Reynisfjara has an appropriately elemental quality for the concluding chapter of my island odyssey. A mist rolls in off the ocean to a melancholy soundtrack: the muffled moans of puffins (*Fratercula arctica*) and the crash of ferocious waves. Just offshore are the imposing Reynisdrangar stacks, the petrified bodies of trolls according to local legend. As so often, the geological and the mythical are dancing together in the inter-tidal zone.

The mist breaks for a moment and I look to the horizon, but beyond the volcanic pillars silhouetted against the sky, there is nothing. If you travel due south from the black beach, my map indicates the next land you reach is the coast of Antarctica, some 15,000 kilometres away. Except, there is something. Scan right slightly and zoom in on the apparently empty waters of the North Atlantic and behold! A tiny teardrop of an island appears marked 'Surtsey' just fifty kilometres away. This is an island younger than me. Its official birthday is 14 November 1963, the day a cook aboard the Icelandic trawler *Ísleifur II* spotted a column of thick black smoke rising from the ocean. When the captain went to investigate, the fishermen became awe-struck spectators as a new island was born. Volcanic contractions rippled up from the mid-Atlantic ridge below, where the earth's Eurasian and North Atlantic plates were slowly moving apart. Eruptions continued for almost four years until the young island fully emerged and could be inked onto the maps, christened Surtsey after a fire giant (*jötunn*) from Norse mythology. There was international excitement at the birth of what is still one of the youngest recognised islands in the world, a rare opportunity for scientists to see how the seeds of life breach the boundaries of barren isolation. Protected by marine exclusion zones, Surtsey was a pristine laboratory, free from human interference, where nature could take its course.

Thawing out in a hotel in Reykjavík, I open my laptop to have a look at what is happening on Surtsey right now. Researchers have placed a live webcam on the island showing a view across what is now grassland, smaller volcanic outcrops visible in the sea beyond. The latest report identifies ninety different bird species which have made Surtsey their home, scores of lichens and fungi have arrived, as well as hundreds of invertebrate species including, to the delight of biologists, an intrepid harvestman spider (*Mitopus morio*). Pangaea seems almost to be squirming on her low bed as I analyse the images and data from the website. The numbers tell me that Surtsey is shrinking, eroded by Atlantic waves. Experts reckon the little island will not survive another hundred years, the sea inexorably reclaiming her territory, just as she had in the Mediterranean when Ferdinandea Island was lost below the waves back in 1831.

Geologically, Iceland itself is one of the youngest islands on the planet, now home to a people who connect with their island status as strongly as anywhere. Isolation has incubated a profound sense of heritage and ancestry that jumps over eight centuries of Norwegian and Danish rule, back to the days when Iceland established the Alþingi (Althing) to control its affairs, the oldest Parliament in the world. Icelanders are passionately protective of the strands of national character that link the present with the days of the Sagas (*Íslendingasögur*), the narratives which chronicle the island's story from the first settlers in the ninth century to the arrival of Christianity 200 years later.

Having only restored the country to full independence in 1944, Icelanders are a people still working out their islandness settings, how open or closed they should be to the world beyond their shoreline. Like the scientists who placed the restrictions upon Surtsey, there are those who believe Iceland should be imposing a cultural exclusion zone around itself, to defend the purity of its way of life. We know, though, that spiders will still find a way to cross the beach and spin their webs.

In the mid-1950s, there was a determined effort to prevent creeping Americanisation from adulterating Icelandic culture. US troops stationed at Keflavík, just fifty kilometres from Reykjavík, were subject to curfew, and local people were prevented from tuning in to American television and radio programmes that were broadcast from the base. It didn't work, of course. Despite a succession of laws requiring television stations to strengthen the Icelandic language and reserve most of their programming for local or European output, Icelanders enthusiastically embraced American TV shows via cable and satellite services. They now lead the world in their use of internet television.

What intrigues me, though, is how alien ideas and be-haviours, which inevitably find their way across an island's boundaries, rarely obliterate what is already there. Instead, they become altered and absorbed into the domestic scene. A short stroll from my hotel, I join the queue of people at *Bæjarins Beztu Pylsur* (The Best Hot Dog in Town), to sample a dish with an American name and culture that is now regarded as Iceland's soul food. Such is the importance of the hot dog in the island's gastronomy that in 2013 a postage stamp was issued illustrating the correct way to serve it. Icelandic hot dogs are made with organic lamb as well as the traditional pork and beef, served with a multitude of toppings, including a particular Icelandic sweet mustard called pylsusinnep and a mayonnaise flavoured with capers and herbs. I am not alone in thinking them a significant improvement on their North American cousin.

Hot dogs in Iceland and the US, like fish and chips in the UK, have their origins in central Europe but were adopted and adapted to become a much-loved constituent of the national cuisine. The aromas and flavours of a culture are infinitely complex, a stew of ingredients and influences that combine and blend to cook up something unique. Cultural purity does not exist, but there is a balance to be struck between welcoming the new and conserving the old. That, as we have seen, is one of the challenges of islandness – working out which strands of national DNA are essential to identity.

For Icelanders, the bloodlines which extend back to the first Viking arrivals are hugely precious. It is a fascination nourished by *Landnámabók* (*The Book of Settlements*), a unique genealogical record, probably compiled in the eleventh century, which details the stories and relationships of more than 3,000 of Iceland's earliest settlers. There are five surviving copies of this national treasure which, along with another medieval text, *Íslendingabók* (*The Book of Icelanders*), still exerts a powerful hold on the country's sense of itself. A database has recently been established with the goal of tracing and recording all known family connections between Icelanders. Much to the amusement of the international press, the developers showed how, if two users tap their phones together, the database app tells them how closely they are related, with an 'incest spoiler' siren if a romantic relationship would be inadvisable.

Until recently, descent from an Icelandic parent was the primary requirement of national citizenship, and foreigners who wished to become citizens were obliged to adopt an Icelandic name. Immigration to Iceland was negligible until it became part of the Schengen Area in 2001, allowing free movement from much of Europe and signalling a big shift in the country's islandness settings. At the time of joining Schengen, one person in forty was a foreign citizen. Twenty years later, the figure had risen to one in seven. There was concern at how this dramatic demographic change would be greeted, but when researchers asked people what impact immigrants had had on Icelandic society, only a handful of Icelanders thought it had been negative, some said it had made no difference, and more than three-quarters of respondents regarded their presence as a positive for the country.

The last two or three decades have been a test of Iceland's self-confidence, a referendum on the country's belief in its capacity to maintain its distinctiveness in the face of global forces. It is a mark of Icelanders' resilience that they somehow weathered the financial crisis that saw their three largest commercial banks default in 2008. They bounced back from the 2010 eruption of the Eyjafjallajökull volcano and the huge cloud of ash which turned day into night and grounded flights across Europe. Indeed, the country decided to commemorate the event by issuing three special postage stamps, the paper of which was impregnated with volcanic ash from the explosions.

I have bought a set of the Eyjafjallajökull stamps and study them under my magnifying glass at the hotel. The 75kr stamp shows a river of volcanic lava and is inscribed with the words 'Bréf Innanlands' ('Local Letters'). The 165kr features the ash cloud billowing from the ice above text which reads 'Bréf til Evrópu' ('Letters to Europe'). The third, priced at 220kr, is illustrated with a dramatic picture of an explosion against the night sky and the words 'Bréf Utan Evrópu' ('Letters Outside Europe'). The three stamps remind me of the concentric circles of belonging that helped me define my identity as a boy, the uncertainties increasing the further one goes from home. It is a quirk of translation, of course, but the word 'Ísland' ('Iceland') on each of the stamps adds to my sense that they are celebrating both isolation and Icelanders' profound relationship with the physical geography of their country.

The bond between islanders and island is personal. Like the stamps, people here seem impregnated by the scoria which bubbles up beneath their feet, emotionally invested in the rock and ice around them. In August 2019, around 100 people, including Iceland's Prime Minister Katrín Jakobsdóttir, walked up Okjökull mountain to attend a funeral service for a glacier. With heads bowed, the mourners listened to eulogies and obituaries for 'Ok', the name given to an ice sheet so emaciated by global warming it had finally stopped moving. A schoolchild read a poem.

> Ok, the burdened glacier
> which at last had had enough
> of acts of terror from men who do not know
> how to have both profits and morals.

'I hope this ceremony will be an inspiration not only to us here in Iceland but also for the rest of the world,' Katrín said, as a brass plaque was fixed onto a boulder, recording the event, and offering 'A letter to the future' (*'Bréf til framtíðarinnar'*).

> This monument is to acknowledge that we know
> what is happening and what needs to be done.
> Only you know if we did it.

Like many islands around the planet, Iceland regards climate change as an immediate and existential challenge. While archipelagos like the Maldives and the Marshall Islands are imminently threatened by rising sea levels, Iceland faces rising land levels. As the glaciers melt, and the weight of the ice lifts, the earth's crust is rebounding, pushing Iceland out of the sea, lifting it by as much as three and a half centimetres a year in some places. The speed of the uplift has startled scientists, who worry the movement and loss of ice will trigger far more aggressive volcanic activity.

I tuck Pangaea into my coat and take the tour bus to Þingvellir (Thingvellir), a national park east of Reykjavík. Þingvellir represents the cultural heart and spiritual soul of Iceland. It is also where I have decided my journey with Pangaea should end.

A thousand summers ago, Iceland's tribal leaders chose this dramatic landscape as the place to establish an island Parliament. The Alþingi first met at Þingvellir in 930 and continued to do so for two weeks each year until 1798. Consensus was sought and found among the deep volcanic fissures that cut through the countryside, the largest of which, *Almannagjá* (Everyman's Gorge), provided the backdrop to the assembly's meetings. Today, an Icelandic flag flies upon *Lögberg* (Law Rock), where for more than three centuries, the *Lögsögumaður* (Law Speaker) recited all the laws of the island from memory, and where, in a remarkable act of unanimity in the year 1000, Iceland peacefully converted to Christianity.

I am sure I can feel Pangaea twitching in my pocket as I walk along Almannagjá, a footpath marking the boundary of two tectonic plates, the high wall of the North Atlantic plate to my left, the Eurasian plate across the valley to my right. Iceland is being ripped apart as the plates separate, while beneath the island a mantle plume, a hotspot of molten rock, surges up to fill the rift valleys and ravines created. Iceland is growing at around two centimetres a year.

Guidebooks explain how the land under my boots is regarded as a bullseye of volcanic activity, the focal point for eruptions and earthquakes which reverberate for thousands of kilometres along the mid-Atlantic ridge. The Iceland plume has had a hand in the creation of a multitude of islands that snake from the Arctic circle down to the Antarctic Ocean: from the doomed Surtsey off Iceland's coast, to the Azores, Madeira and the Canary Islands, onwards to Bermuda, Ascension, St Helena and Tristan da Cunha, a string of insularity which eventually reaches *Bouvetøya* (Bouvet Island), a Norwegian nature reserve in the South Atlantic that is more than 1,000 kilometres from anywhere else and regarded as the most isolated place on the planet.

I take Pangaea from my jacket and sit with her beside the basalt cliff of Almannagjá. I cannot help feeling that we are perched on the edge of the world, which, in a way, we are. Far below us, unseen, grinds the engine which powers the planet, the source of the geothermal energy that moves continents and shapes the oceans. We are on the shoreline of one of the great shifting landmasses as it slowly wanders across the surface of the earth, its geological partner inching away in a tortured boléro. Hot geysers, which regularly burst from the ground in Iceland, put on a show of steam and spray to remind islanders of the forces at work beneath them, evidence of an alternative history in which time is measured, not in minutes or months or millennia, but in epochs and eras and eons.

I cup Pangaea in my hands and consider the original island after which she is named, the supercontinent which began to break apart 200 million years ago, its constituents drifting into the positions so familiar to us today, islands great and small. Geologists now believe the earth's building blocks are on a journey of reunification, a process that will, in time, create a new supercontinent as the plates huddle together once again. This hypothetical landmass has been given various names: Pangaea Ultima (the final Pangaea), Pangaea Proxima (the next Pangaea) or Novopangaea (the new Pangaea). Many scenarios have been suggested, but there is general agreement that the earth's great engine powers a sequence of tectonic movement, over hundreds of millions of years, in which a supercontinent breaks apart and then comes back together again, completing the eternal circle.

I cannot help but feel moved by this story. It is as though the earth is scattering its seeds and then harvesting their bounty, a cycle of dispersal and gathering, of isolation and connection. It is the Alpha and the Omega.

I look at my sleeping Pangaea, this cheap souvenir from a museum gift shop, and wonder at how she has led me on a journey of enlightenment and discovery that I know must end here. I had hoped that, with her help, island shorelines might somehow lead me a little closer to the truth. In fact, I feel both nearer and further away, with the realisation that our inadequate brains cannot hope to comprehend even a fraction of the complexity of the universe. We have almost as little understanding of what holds true about our world as a piece of driftwood knows of the beach where it washes up.

If I have learned anything of value it is this: meaning is found in the margins, at the points of connection and contradiction, where 'us' meets 'them'. Island syndrome shapes us all. Fulfilment requires us to step across the threshold and beyond our shoreline, to search for answers and to take risks. We know we can never reach the truth, but we must forever seek it. That is humanity's burden and its blessing.

I stand up, a light rain against my face, and place Pangaea in an inconspicuous crevice in the great wall of Almannagjá. It is time for me to head back to harbour, to take the boat to my Ithaca, the island where I belong.

ACKNOWLEDGEMENTS

This book's gestation has been longer than the pregnancy of a frilled shark (*Chlamydoselachus anguineus*) cruising off the beaches of the Azores. More than forty months, if you were wondering. In that time, I have been blessed with amazing support from family and friends. Huge thanks are due to my wonderful wife, Antonia, who cast the wisest of eyes over my words, and to my children – Flora, Eliza, Annis and Ed – whose invaluable perspective and various knowledge sets make me proud to be their dad. I would also like to thank my brother Robert, my friend John Kampfner, my agent Andrew Gordon, my editor Ella Boardman and all the terrific people at Biteback for advice, support and expertise.

INDEX